Keeping goldfish - Guide to keeping goldfish in the pond or aquarium

Buying , keeping, breeding goldfish, indoors or outdoors - info on keeping, environment, diseases and diet

AF884716

Publishing Expert Group

Keeping goldfish - Guide to goldfish in the pond or aquarium

Buying, keeping, breeding goldfish indoors or outdoors - info on keeping, environment, diseases and diet

Publishing Expert Group

TABLE OF CONTENTS

ABOUT THE AUTHOR

Svenja Nett lives with her husband in a small village near Aurich on the German North Sea coast.

As a zoologist, she was involved with pets of all kinds from an early age and gradually specialized more and more in aquatic life and here in particular in ornamental fish. For more than 20 years she has been running a successful aquaristic shop and advises aquarium and pond owners there or directly at their homes. Over the years, she has acquired extensive knowledge on many aspects of keeping fish and other aquatic creatures.

The idea for this book came about in order to be able to pass on this knowledge to more than just a small circle of local aquarium enthusiasts. It should give all prospective or existing owners of aquariums or ponds an introduction to the world of goldfish keeping. Keeping goldfish is not difficult in principle, and yet newcomers to this field in particular often make mistakes that usually cost money, effort and fish lives.

Svenja Nett would like to give every interested person the necessary knowledge to do everything right from the beginning and to offer the fish the habitat they need and deserve. Then you will have immediate success in keeping your fish and unnecessary effort and frustration is avoided.

Each of her publications is therefore based not only on the indisputable scientific foundations, but also on very personal experiences and insights. This way, the reader not only receives factual information about the subject but also a practical guide with broad knowledge and useful advice that can be easily understood and implemented.

FOREWORD

Goldfish - a beautiful hobby for lovers as an eye-catcher in the aquarium or garden pond, which some people can only dream about. Others think that the beautiful ornamental fish, after all, do nothing but swim in circles, eat and defecate.

A pet, which is not much work, many think. But what really needs to be taken into account? What does a goldfish need at all? How high are the costs for purchase, care and does a goldfish actually need a vet?

These questions and many more are answered in this guide. In addition, you will receive all kinds of information about the fish, which is more versatile than many people think at first glance. As with humans, there are some similarities in the appearance, size and origin of goldfish, but there are also many differences. You will learn more about this in the following pages.

It is considered a fish for beginners and children. This is true if we only buy the simple familiar goldfish type. Temperature, oxygen content of the water, food - the simple goldfish is not picky, we can even say that it is tough. It has inherited all these peculiarities from his ancestors - crucians.

With goldfish, two characteristics are particularly noticeable, which make the differences to other beginner fish clear. First, there is the longevity. Since it is so robust, it can be assumed that it will live for many years. Secondly, there is its size to consider. It is indeed the case that goldfish can grow quite large. For this reason, coupled with its strong desire to socialize, you will need to create a large aquarium or pond.

As you can see, there is a lot to discover and get to know. Accordingly, the goldfish is not a boring old friend, but can be quite exciting. What are you waiting for? Keep turning the pages and see what there is to learn.

- Chapter 1 -

ORIGIN OF THE GOLDFISH

THE STORY OF THE GOLDFISH

The goldfish is a vertebrate. Some of these vertebrates have mutated, as did the ancestors of the goldfish. In **some of** them a melamine deficiency occurred, which is why the upper pigments, which are actually dark, were missing. Below these pigments are carotenes and pethidine, which are responsible for the orange coloration of the fish.

The first reports date back to the 3rd century BC. According to these, fishermen discovered orange-colored fish, which was unusual for its time. In those days, people mainly knew the silver crucians, the predecessors of the goldfish. Because of the peculiar coloration, the orange fish were not considered edible.

Buddhism spread in China between the 5th and 7th century. The main principle of the Buddhist teachings – the avoidance of suffering – is still as valid today as it was then. The orange fish fitted well into those surroundings and became greatly admired. There are numerous temples in China, near which ponds were built at that time, also called grace ponds. The beautiful fish were placed there. Turtles also found a new home there.

Emperor Song Gaozong ruled the country from 960 to 1279. At that time there were numerous of these grace ponds, where the colored crucians multiplied. Thus, the xanthism, that is, the orange discoloration, was inherited. In this way, the stocks of orange fish became larger.

West Lake in Hangzhou is another place where the goldfish were discovered. The emperor was asked to declare it a mercy lake and as a result, numerous fish were released into it. From that time onward, Hangzhou has been considered the place of origin of the goldfish.

Did you know? Among Buddhist monks, goldfish were also considered celestial miracles! The emperor himself also kept goldfish. For him they were ornamental fish and were used for representation. Among the nobility, the beautiful fish were revered. Only expert personnel were allowed to take care of their reproduction and care. This was thus the birth of the profession of goldfish breeder.

Of course, the result was that the keeping of goldfish spread all over China. During the Song Dynasty, the areas of Jiaxing, Hangzhou and Nanjing were especially considered the main places of goldfish keeping. However, they also came to the north of China where it was a little cooler. Around 1330, goldfish also reached Zhenjiang and Beijing during the Yuan Dynasty. At that time, the non-noble population also kept the beautiful ornamental fish. They kept them in ponds and continued to breed them. Since then, other color varieties have been found among the breeding stock. The multicolored goldfish became particularly popular.

During the Ming Dynasty, from 1368 to 1644, attitudes towards the goldfish began to change. Miniature ponds were created in large tubs, where the goldfish and other animals were placed. Later, the tanks became even smaller and even the families who had very little money could afford a goldfish. The so-called dragon bowls, vessels made of wood or baked clay, were created. Families who lived in prosperity had such bowls made of porcelain.

Mutations of fish that would have had no chance to survive in the wild were kept in these vessels. Interest in these and in goldfish in general was very high at the time, so people did not have to ask themselves whether or not they should actually continue breeding.

In 1569, the first book about goldfish breeding was written. Zhang Qian De described the different color varieties, fish with bulging eyes, dragon fish and some more.

Towards the end of the Ming Dynasty, there was hardly a home in the empire that did not have goldfish. Almost all people were enthusiastic about the goldfish. It is also interesting that within just 100 years most of the still known pet fish were bred.

THE GOLDFISH IN EUROPE

The magnificent looking Goldfish of today did not occur naturally, but were artificially bred. Goldfish belong to the carp family, and are originally descended from the Prussian carp. Carp are most common in Europe, Asia, Africa and North America, and are a freshwater fish.

If we look at the ancestors of the goldfish, we see no similarity in color, because the Prussian carp are dark greenish brown to blackish gray. Silvery stripes decorate their sides and give the fish a beautiful sheen. Goldfish are yellowish to red-orange, usually with white or black spots. Unlike the Prussian Carp, they shine golden.

The special thing about today's goldfish is that they come in numerous colors and breeding forms. Here, especially their body shape and fins play a significant role. Even the eyes can be impressive, especially in the species that have spherical eyes that protrude.

In Southern Europe there are also feral goldfish which have found a new home in rivers or ponds. As already described, China was the first country to breed goldfish. Therefore, it can be assumed that the goldfish found its origin there.

If we look at the origin of Prussian carp, this actually fits, because they come from East Asia. Moreover, it is said that goldfish were living in China 2500 years ago. Goldfish live in freshwater and love tranquility and lots of plants around them.

As already mentioned, the known ancestors of the goldfish are the Prussian carp. In China at that time, they were considered sacred if they had a red-gold coloration. The original Prussian carp were considered edible fish in China and were eaten with pleasure. This was not the case with the goldfish.

But when did the goldfish come to Europe? It was only between 1611 and 1691 that the popular ornamental fish finally made it to Europe. Mostly it was found in England, Portugal or the Netherlands. At the beginning of the 18th century it did not take long in England before the golden fish had gained popularity. Madame Pompadour (mistress of the French King Louis XV) was supposedly given some goldfish as a gift. Based on this, it was then able to establish itself throughout Europe.

In 1797, Bechtstein wrote the "Naturgeschichte der Stuben thiere" which roughly translated means “Natural History of the house pet”. In it, he mentioned numerous animals, but only two species of fish. One of them was the goldfish. At that time, only the extremely hardy species of goldfish could be kept in the house. At that time, little was known about how to take care of them.

The goldfish spread in Europe during the rococo and chinoiserie periods. These ornamental fish from China fitted well into the times. One thing, however, did not find favor here and that was the dragon bowl made of porcelain. Actually, this is surprising, because Chinese porcelain was basically very popular at that time.

Instead of dragon bowls, people offered the goldfish a new home in round glass bowls, which, as we fortunately know today, were not exactly suitable for keeping goldfish, because they held far too little water and offered the animal hardly any space.

Around 1872, the beautiful veiltails were imported from Japan to Paris. In other areas of Europe, too, people now ventured into the more difficult breeding forms, first and foremost in Germany. Paul Nitsche (chairman of an aquarium club) and Paul Matte (ornamental fish breeder) had a real competition in goldfish breeding at that time. Ernst Bade also wanted to get in on the action. He contributed to the legend of Matte's breeding methods.

- Chapter 2 -

ABOUT THE GOLDFISH

THE APPEARANCE

Many will think: The name already says it! A goldfish is gold colored. But there I must disappoint you. Most species of goldfish have a yellow-orange or even orange-red coloration. But - and now it gets interesting - there are even white goldfish, which have black spots.

Like all fish, the goldfish has a body - consisting of a head, trunk and caudal peduncle - a tail fin, a dorsal fin, two pelvic fins, two pectoral fins and an anal fin. Let's take a closer look at the different fins: Basically, they are all fins, but these show differences and not only in function. For example, they can be single, but also in pairs.

Each individual fin is crossed by so-called fin rays. These are nothing more than skeletal components, i.e. bones. Likewise, fins use various muscles, which are responsible for the erection and folding of the fins. The muscles also enable the fish to move through the water or to stabilize themselves in it.

Now let's take a closer look at the different fins:

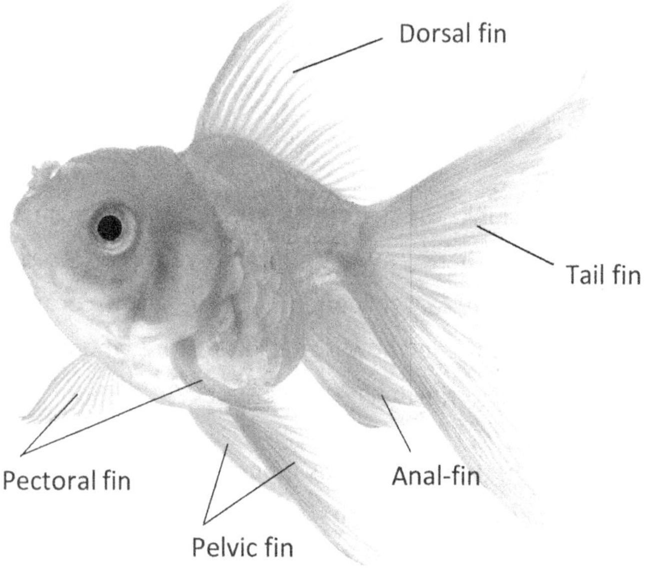

The tail fin

In Latin, this fin is also called caudal. With this fin, the fish makes sinuous movements and is irreplaceable for its locomotion. The reason for this is that it can exert so much force that the fish can actually move.

Some goldfish species have a special tail fin, because although it is simple, it can easily be assumed to be paired. Some goldfish have a split tail fin, which looks like a veil.

Unfortunately, the split veil fins don't do the fish much good because they don't give it enough propulsion, making it very slow to move in the water. If such a fish lived in the wild - that is, in the sea - it would often be helpless against its enemy. The fish's body and caudal peduncle must make strong sinuous movements to enable it to move reasonably quickly through the water.

The dorsal fin

It has the Latin name dorsal. Its function is to stabilize when swimming. We could also say that the dorsal fin is the steering wheel of the fish, as it prevents it from tipping over as soon as it takes a turn, for example.

Dorsal fins have different types of fin rays. A distinction is made between soft rays and hard rays. Hard rays are unarticulated, usually smooth pieces of bone, soft rays consist of two halves fused together. The types can also be used to determine the species of fish. True hard rays are only found in the spiny fins.

The anal fin

It is also called the anal fin or anal. Its function is similar to that of the dorsal fin: it stabilizes. In female goldfish it is thickened, but this can only be determined if the fish are compared directly with one another.

The pectoral and pelvic fins

They are also called pectorals and ventrals. These fins are used for steering. Sometimes, especially when the fish are in still waters, they swim along quietly. These movements are also controlled by the pectoral and pelvic fins. They also help the fish to maintain the calm position while they are burrowing (i.e. taking food from the bottom of the water).

The body of these goldfish is high-backed (they have a kind of hump). In addition, they have an underslung mouth. What is that? Let me explain it briefly: In most fish, the mouth sits in the middle of the head. However, in goldfish, it is located on the lower half of the head.

If you want to distinguish between males and females , you need to look at the fish quite closely, otherwise you will not notice the subtle differences. In fact, we cannot recognize the differences between males and females in a goldfish until they are fully grown, at least one year old. As a rule, females are smaller and thicker than their male counterparts. The latter are long and slender.

In addition, the cloaca indicates whether the goldfish is female or male. While it protrudes in females, it is curved inward, elongated and narrow in goldfish males.

SKELETON AND INTERNAL ORGANS

You have now become acquainted with the outer form of the goldfish. At this point, let's take a look at its interior.

The goldfish also has a spinal column. This carries its skull, in which its brain is located. The ribs and bones are responsible for the protection of the internal organs. It also has a jaw, but without teeth. Instead, there are four teeth at the end of the oral cavity on each of the gill arches, which are used for grinding food. As a counterpart to these, the fish carries a bone plate on the upper side at the gullet bones, which is horny.

The goldfish does not have a stomach. The esophagus opens directly into the midgut. Here we also find the excretory ducts of the liver and pancreas. These release the digestive juices. The nutrients from the food are absorbed directly into the blood via the intestinal mucosa.

The swim bladder with two chambers emerges from an outpouching of the foregut. If we look at the history of development, this has the same origin as the lung.

The lungs are used for breathing, but the swim bladder is not. This is a so-called hydrostatic organ. It releases and absorbs gases. Accordingly, it is responsible for the buoyancy of the fish. This allows it to float in different depths of water. The shape of the swimming bubble, as well as its position, are also responsible for the position of the fish. We can also compare it with the eardrum, because it also picks up sound waves. The vibrations are then transmitted through the Weber ossicles. Did you know that goldfish can even hear very well?

A goldfish also has kidneys. These even have a lot to do. They regulate the water balance and ensure that a fish living in seawater develops a high fluid requirement and a fish living in freshwater loses the excess fluid. In freshwater fish such as the goldfish, the fish's tissues have a very high salt content. For this reason, water constantly flows into the fish, which must be excreted again. Therefore, if the kidneys are severely damaged, the fish can swell up considerably.

The goldfish has two types of arteries. The afferent arteries of the fish are responsible for carrying blood from the heart into the gills. There, it absorbs oxygen from the water and releases carbon dioxide. In addition, the goldfish also has efferent arteries. These carry the blood from the gills to all parts of the body. The cells now consume the oxygen and produce carbon dioxide. Veins lead the blood back to the heart. So you see: this circulation is completely different from the human circulation.

The fish also has a lateral line organ, or in other words, tubes equipped with hair cells. Their task is to perceive water movements and the pressure differences as well as deep sounds. From this, the similarly constructed inner ear was developed.

The physique of the different fish varies tremendously, according to their breed. Some only show differences in coloring, or size and shape of their fins. However, there are also breeds , in which the torso or also the head show a clear thickening. In some of them the caudal (the tail fin) has doubled or has split. Also a possibility is the absence of the dorsal fin. Now and then only small remnants of the dorsal fin remain. This is not intentionally bred and hardly pleases anyone.

Some breeding forms have a cap on their head, which consists of fatty tissue. Normal breeds ings also show this, but in certain breeds it is massively thickened.

Then there are breeds in which the eyes are strongly enlarged and protruding. Due to the over production of thyroid hormone, a thickened retina occurs , which is the cause of these peculiar eye shapes. Fish, which have irregular connective tissue, can show upturned eyes like the Stargazer.

The bubble eye goldfish has upturned eyes, under which are bubbles filled with lymph. Others have so-called pompons, enlarged, fleshy nasal humps. In rare cases, goldfish breeders also have curled gills. These are gills that are curved forward. Other breeding forms have hemispherical shaped pearl scales, caused by lime deposits.

A typical domestication feature is also a shortened facial skull and displacement below the cranium. This is the case with Wakins, for example. These fish can look very pug-headed depending on the breeding.

There are Ryunkin forms in which the facial skull is reduced in size and pointed. The head resembles that of a mouse. Normal goldfish have no ridges on the top of the skull, while in others we see an enlarged head outgrowth.

Another example is the more curved spine, which can occur in some compressed breeds. However, this is very rarely the case. In normal goldfish the spine is curved differently. In high breeds , the spine may be somewhat shortened, but the gut may be elongated. In stocky breeds , the caudal chamber of the swim bladder is much smaller. Sometimes it simply has a different position or is missing completely.

As you can see, there are many different forms here, which are due to the breeding that took place a long time ago. Over the course of the centuries, more and more variations have developed.

THE SIZE

How big a goldfish grows depends on where it is kept. Goldfish, which live in an aquarium, usually remain only a few centimeters in size. Others, however, which can be found in garden ponds or even in larger aquariums such as zoos, can grow up to twenty centimeters. If the fish have a lot of space available, they can even reach a considerable size of up to fifty centimeters.

SPECIAL FEATURES

Like all fish, the goldfish has a bony skeleton.

Goldfish belong to the carp family which do not have stomachs and, as previously mentioned, this beautiful fish must also do without teeth. Instead, they have strong gullet teeth, which are located in the throat. Goldfish take their food through the mouth, which they can protrude. The food is then crushed with the gullet teeth and the carp stone (as the chewing plate is called). Since the goldfish has no stomach, the food is digested in the intestine.

Goldfish have scales all over their body, even on their heads. However, here is a breeding form that has no scales on its head. This is called oranda.

In addition, goldfish have a lateral line that is fully equipped and visible. The lateral line is the externally visible part of the lateral line organ. It is used to record water vibrations and thus enables the fish to obtain accurate information about movements in the water.

Even the goldfish's eyes are special. They have four different types of cones, one more than we humans. In addition to the familiar three cones for red, green and blue, they have a fourth, the UV cone. This picks up short-wave, ultraviolet light and can even process this information. In short, goldfish use four color receptors with which they can see.

If you wonder how a goldfish hears, you will quickly be amazed, because the small fish hear almost at the same frequency range as we humans. The so-called swim bladder picks up the sound waves and passes them on to the inner ear.

BEHAVIOR

If you want to get goldfish, you will certainly be interested in their behavior, to get to know them properly.

Goldfish are very greedy, so it is very likely that they will beg from you regularly. In addition, goldfish are very smart and learn quickly. For example, when you approach the aquarium, the fish will immediately swim to the place where you have always given them food.

Do not be fooled into thinking that your fish are hungry at that very moment. The rule of thumb is: If the food has not been completely eaten within two minutes, your fish are not hungry, despite their apparent appetite. It is better not to give them too many snacks and to let them go hungry on occasions. Once a week, you could forego feeding them, so that they have to search for leftovers in the water and that also protects them from getting fat.

Goldfish love to dig, but unfortunately this does not always look very pretty. By digging in the ground, all sorts of dirt is stirred up, which makes for cloudy water. Therefore, it is advisable to buy a filter that has a very strong performance.

The negative behavior of the goldfish also includes that they eat their own spawn. However, since a female lays up to 3,000 eggs, there is still a lot of offspring left over. Since the young fish are still black at first, they have to earn their name "gold fish" over time. They only acquire their red-orange color later.

Goldfish can cope with only a small amount of oxygen in the water. The reason for this is their gills, which are well supplied with blood. Nevertheless, it is important that at higher temperatures - i.e. above 25 degrees Celsius - you top up with cool water every now and then and that the circulation pump remains in constant use. This does not apply to the aquarium, where temperatures do not vary.

All in all, the goldfish is rightly a very popular ornamental fish. It is easy to care for and beautiful to look at. With its open nature and its ability to learn, it will give you a lot of pleasure.

FRIENDS AND ENEMIES

Every animal has natural friends in the wild - but also enemies! So does our dear little goldfish. If kept in a garden pond or aquarium, it is rather rare that they have enemies. But unfortunately it happens again and again that people abandon the fish, for whatever reason. Therefore, you should also consider very carefully whether you really want to keep goldfish.

Goldfish, which were abandoned, become feral, and unfortunately, therefore, they do not know who their enemies are. They trust virtually all creatures that they encounter. But their striking color becomes their undoing. They cannot hide well and their enemies can find them easily. In addition, they become popular prey. Not only predators hunt the small goldfish across the water, but also birds of prey, since they can spot the fish from a few meters above the ground.

For more info on goldfish friends, see the later chapter on goldfish companions.

- Chapter 3 -

THE LIFE OF THE GOLDFISH

HOW DOES THE GOLDFISH LIVE?

Goldfish are sociable swimmers. They only feel really comfortable when they live in a small school. This is exactly why these beautiful fish need space.

When they are small, it is quite possible to keep them in a normal aquarium. However, once they are fully grown, they are much more comfortable in a pond, where they can really spread out. If you are considering actually getting a goldfish, you should think about buying a small school, which would best be kept in a pond or an exceedingly large aquarium. In addition, you should also know that goldfish reproduce abundantly outdoors.

Goldfish are considered peaceful fish, which swim quietly around the pond. However, they also like to dig in the ground. So, if at some point you discover dislodged plants in your pond, it could be that your goldfish were responsible.

THE DIFFERENT FORMS

The goldfish is an artificially bred fish. Today we know that there are well over 120 different forms of this fish species.

An example is the veiltail, which, as its name suggests, has a veil-like tail. The fantail is similar, in that its tail fin resembles a veil, but looks much more like a fan.

Other forms are, for example, the lionhead. Its peculiarity is the head, which has strange protrusions, the pearlscale, whose scales are arched up or also the stargazer, whose eyes are directed upwards and protruding.

Other well-known and popular species of goldfish:

- Shubunkin
- Wakin
- Jikin (peacock tail)
- Watonai (Waryu)
- Kinranshi (Clouded Eye)
- Sarasa
- Shukin
- Black Moor
- Telescope eye
- Ryukin
- Tancho
- Demekin

- Guckyn
- Hanafusa (nasal bouquet)
- Nankin
- Osaka Ranchu (Osaka Buffalo Head)
- Eggfish
- Ranchu (buffalo head)
- Celestial
- Pompom
- Bubble eye (large bubbles)
- Toad eye (small bubbles)
- Meteor
- Tosakin (Tosa)

LIFETIME

If you want to get goldfish, it is important to also consider their life expectancy. After all, goldfish can live up to thirty years. A considerable age for such a fish. But the fact is also that in rare cases they can live even around forty years.

How old a goldfish can actually become depends not only on its physical condition, but also on the conditions in which it lives. An owner who cares for his fish and is always concerned about their well-being can certainly expect his goldfish to reach a ripe old age.

If you ask around and research how old the high-breds can become, you will unfortunately often read that they die early, if you find anything at all about it. So-called high breeding focuses on a targeted selection of parents and selection of offspring for certain characteristics, through which the quality of the fish increases. Unfortunately, I must say at this point that it makes a big difference how well the fish are looked after. High breeding is not suitable for beginners. Unfortunately, people often buy them and are ultimately overwhelmed when something goes wrong.

- Chapter 4 -

THE REPRODUCTION

Goldfish spawn from April to May. Breeding resembles a drive hunt, because the male chases his beloved across the water. In the process, they repeatedly bump into each other. This stimulates the female to lay eggs. After a certain time, the females are exhausted, which we cannot blame them for. They then sink to the ground. And as if the male then wants to make up for his actions, he carries his female up to the surface by her snout. There the female lays between 500 and 3,000 eggs. These are then fertilized by the male.

After five to seven days, the time has come. The larvae begin to hatch. These are almost transparent and very sticky. Therefore, they initially stick to the water plants. So if you create a garden pond and want your goldfish to reproduce, you should definitely take into account that not only is there enough space, but also that the fish have enough plants in their pond.

As they grow, the tiny, almost transparent larvae become known as fry. These feed on microorganisms that are found in the water.

Initially, the little fish are dark gray. They change color only after ten to twelve months as follows: From dark gray they change to black. After some time, their belly begins to turn golden yellow, until they finally get their red-orange color. In the last step of the coloring process, the white spots appear.

CONDITIONS FOR REPRODUCTION

In order for goldfish to reproduce, certain conditions must be met. I would now like to briefly explain these to you:

1) The pond or aquarium must be large enough for the goldfish to perform their spawning chase.
2) There should be different levels in the pond. Here it is important that the pond has a depth of one meter in at least one place. They serve as comfort zones for the fish.
3) Sufficient aquatic plants are necessary for the larvae to attach to them. In addition, the plants protect the young fish.
4) The temperature of the water must be at least 15 degrees Celsius. Before that female goldfish do not begin to mate. The reason for this is that the rising temperatures make the belly of the female increase in size, so that there is more room for the eggs.
   In summer, the temperature of the water should be from 18 to 24 degrees Celsius. Therefore, it is important that half of your pond is in the shade.
   In winter, the temperature should not be lower than six degrees Celsius. Here,

however, the respective breeding form must be known exactly, since some also require somewhat warmer temperatures in the winter.

You can ensure the right temperature if the pond is not too shallow.

Goldfish are very hardy fish, if we take a closer look at the quality of the water. And yet they feel more comfortable when the water has a certain quality.

Tip: Before introducing the fish, you should carry out an analysis of the water or have it carried out by a specialist.

If you decide to do this analysis yourself, you should use the following values as a guide:

- PH value: 7 to 8
- Water hardness: 12 to 18 dH (degrees of German hardness)
- Carbonate hardness: 10 to 14 dH
- Less than 0.1 milligrams ammonium per liter
- Less than 0.1 milligrams of nitrite per liter
- Less than 25 milligrams of nitrate per liter

- Less than 20 milligrams of carbon dioxide per liter

The water quality must always be satisfactory, otherwise the fish do not feel well or can also fall ill.

5) The bottom of the pond / aquarium must be as natural as possible, because they are digging.
6) As already mentioned, goldfish dig for food. In doing so, they ingest aquatic organisms and plants as their food. If you set up the pond in a species-appropriate way, you do not have to feed them every day. However, it also depends on the number of fish. If the number is high, it is necessary to feed more. If you add feed, you should always buy it in a specialized store. This way you are on the safe side. Usually the fish are given pellets, flakes or granules. However, you can also give them a little pleasure with some variety. Mosquito larvae and water fleas, for example, are wonderfully suitable for this.

   Important! Make sure that no remains are left in the pond, as this massively damages the water quality. It is therefore advisable to carry out feeding in several stages. If food remains after two minutes, this is already considered as leftover food.

The right conditions are one thing. However, it is much more important that the fish are also willing to reproduce. This is where your willingness to observe is required.

Once the male starts nudging his female constantly, this is a sign that they are ready.

THE SPAWN

The spawn of the goldfish is almost transparent (it has a silvery color) and is very small. Very small means that it is only a few millimeters in size. Once the female has laid the eggs, they are fertilized by the male in the water. The spawn then attaches itself to the plants in the water.

The larvae are surrounded by a so-called yolk sac. From this they feed during the first days. Once they have hatched, they ingest the water organisms.

It is also interesting that the female lays up to 3000 eggs and this can happen up to ten times during the spawning season. If you now add up the numbers, you will be frightened and think to yourself: Lord, where am I going to go with all these fish? Here I can quickly reassure you. Not all the eggs will hatch into fish. Some will not be fertilized and again some will already be eaten by the adult fish themselves.

PROTECTIVE MEASURES

Anyone who wants to try breeding, must first and foremost ensure that the spawn is protected, especially from its voracious enemies. Not only other fish that may swim with in the pond or aquarium, but also other animals that consume the spawn from the water surface, are among the enemies here. In fact, it is primarily the goldfish themselves that regard the spawn as their food. In short, the beautiful ornamental fish are cannibals.

The fry are just two millimeters in size after hatching. At first they have a dark color, which is their natural protection. Between all the water plants they are difficult to see.

If you want to try breeding yourself, it is advisable to create the pond on several levels. These serve not only as oases of well-being for the adult fish, but also as a shelter for the young. It is best to plant a shallow waterhole with plants that are as dense as possible. There the young fish can hide wonderfully, since they do not stand out under the dense plants with their dark color and the adult fish are not able to swim well in shallow water.

Another protective measure would be to fish out the spawn. You can fish it out immediately after fertilization and place it in an aquarium where the young fish can grow up protected. Here, however, endurance and patience are required, because you need to watch the fish closely so that you do not miss the spawning time and fertilization.

It is important that you design the aquarium species-appropriate, because the young fish spend their first year of life in it before you put them in the pond or give them to interested people.

An aquarium always has a suction tube so that the water is always the best quality and the oxygen content is consistent. However, this is very dangerous for the fry, which are only a few millimeters in size. It is recommended to cover it with fleece, so that the small fish do not get sucked into it.

- Chapter 5 -

TAKING CARE OF GOLDFISH

Simple goldfish species are not very demanding and therefore easy to keep. However, this is different with breeding fish, as some can be very vulnerable. Therefore, only people who have already gained experience with breeding fish should also keep them.

First and foremost, the fish need space, whether simple goldfish or special varieties. Therefore, the pond basin, which you choose, should be at least 1 meter long. But even better, in fact, are those which significantly exceed this size. I usually calculate with three liters of water per centimeter of fish. Accordingly, a ten centimeter long fish needs thirty liters of water.

However, a goldfish should not be kept alone, as it loves to be around its own kind. You should definitely purchase your goldfish from a professional, as he will be able to explain everything you need to know. This person will also tell you that at least three goldfish should be kept together. In this case, it is important that you make sure to buy a pond tank that can hold at least ninety liters of water.

In addition, the tank should have different levels, because goldfish also love variety and the so-called comfort zones between the plants, where they can retreat. Refrain from using small, round bowls. These are not species-appropriate, but border on animal cruelty. Also an aquarium is not a species fair place to keep the beautiful ornamental fish, unless you have so much space that this can be larger than average. Keeping them in a pond is therefore species-appropriate and optimal.

A pond or basin does not require heating. If you create a pond, which is at least one meter deep, the goldfish can stay outside even in winter. Then the water does not freeze to the bottom of the pond / basin.

Whether you choose a pond or an aquarium: It is important that there is a bottom layer of gravel. In addition, as many aquatic plants as possible should be planted.

Of course, you can also decorate the pond or aquarium. For this purpose, large stones, parts of roots or even the coconut shells are wonderful.

WHAT DO GOLDFISH EAT?

The goldfish is the pig among fish. Not because of its appearance or behavior, no. It eats everything. Fish food, plants, water fleas, worms, mosquito larvae, but also oatmeal, egg yolk and corn are often on its menu. But it is not only important to know what can be fed, but also how often and how much should be fed.

Many people who would like to have goldfish, fear the rumor that goldfish may starve over the winter. However, this is not the case, because the goldfish stay near the bottom and fall into winter torpor. In this state, they do not need food.

Goldfish can beg! And if you now think that this is not true, you are wrong. The fish notice exactly when we are near the pond and with time they get to know us well. They know that humans bring them food and treats. They quickly swim to the immediate vicinity of the water surface and adore their owner.

And as we humans are, we tend to give them treats. Unfortunately, this little snack is often too much, especially if we give it to the fish more than once or twice.

Give the adult fish in your pond fish food once or twice a day. You can give twice the amount to fish which are still growing.

The quality of the water also plays an important role in feeding. If you notice that the quality is deteriorating, be sure to reduce the amount of feed, because food residues cause contamination of the water.

You should also always be careful not to feed too much, and watch the food intake of the fish. The food should already be eaten within a few minutes. If any remains, you must remove it so that the quality of the water does not deteriorate.

What about when you go on vacation and no one can take care of the fish? Many beginners often worry that the fish may starve during this time. But in fact, fish can last a whole two weeks without supplementary feeding from humans. However, the prerequisite for this is that the pond offers them enough aquatic plants, because they are also eaten by these beautiful fish. In addition, they also feed on aquatic organisms.

If you have had a pond with fish for only a few months, you can still go on vacation with peace of mind. The fish will still find something to eat.

Anyone who gets goldfish must of course also deal with the question of whether goldfish are allowed to eat bread. It is not always easy to feed your pet properly. Some do not like the special food, others may have allergies. Yes, there is such a thing! Goldfish, for example, love to eat bread. But is that at all healthy for these beautiful ornamental fish?

Bread and other treats contain carbohydrates , which are transformed in the digestive tract. Gases are formed, which can affect the function of the swim bladder, but unfortunately negatively. Bread does not immediately ensure the death of the fish, but it is just not healthy.

Another negative aspect of bread feeding is that the water suffers because algae formation is promoted. Moreover, the fact is that at least goldfish living in ponds usually find enough food, so we do not really need to feed them. Those who have doubts about this and think that their little fish might starve can give them fish food every few days.

Those who want to feed their fish bread should get into their heads that it is given a) in small measures and b) only very rarely! However, it is not advisable.

VITAMINS FOR GOLDFISH

Many people take additional supplements to maintain their vitamin balance. But do fish also need additional vitamins? Let's take a look at what arguments might support this:

- Vitamins increase the colorfulness
- In the water the fish lose vitamins through their gills
- Many vitamins that are in the feed are destroyed during packaging (by high heat)
- Many vitamins oxidize at high temperatures
- With too much light some become unusable
- The content of vitamins in fish food could be insufficient
- Frozen food can be improved by adding vitamins
- Vitamin deficiency can be difficult to detect in fish

These arguments speak against it:

- Varied, fresh food: Giving vitamins not necessary

These foods are natural and recommended for fish:

- Vegetables

- Fruit
- Pond food (live)

When is it useful to give the goldfish additional vitamins?

- With fish feed that has been stored for several months
- If he dry food has already been open for several months
- For stressed or sick fish
- In the spawning period

What vitamins are necessary?

- Vitamin A for vision
- Vitamin D for "bone formation" in young fish
- Vitamin E for the regulation of aerobic metabolism
- Vitamin K for the coagulation of blood
- Vitamin B1 for the enzymatic substance exchange
- Vitamin B2 for metabolism (carbohydrates)
- Vitamin B6 for metabolism (protein and fat)
- Vitamin B12 for the formation of red blood corpuscles
- Vitamin C for growth

How to give vitamins to the goldfish?

Vitamins are administered with food, that is quite clear. But fish are also capable of producing vitamins themselves.

As you may know, there are water-soluble and non-water-soluble vitamins. Goldfish need both categories. If we add water-soluble vitamins to the water, the fish will easily absorb them through their gills. It is important that the concentration of vitamins is high. Unfortunately, many vitamins are also quickly lost because not all of them can be absorbed. Fish cannot absorb vitamins at all that are insoluble in water, but are simply added to the water.

The best way is to mix them into the feed. There are pastes that adhere very well to the food. For example, goldfish love water fleas. When they eat them, they automatically ingest a lot of vitamins. It may sound a little unattractive, but the stomach content of these insects is even more nutritious than the insect itself.

How many vitamins are needed by goldfish?

Right at the beginning you should know that an overdose of vitamins can harm the fish just as much as a lack of vitamins. For example, metabolic functions or the work of the kidneys could be impaired. Therefore, it is quite sufficient to give vitamins to the fish twice a week. If you are feeding fresh food, they are already receiving enough. It is therefore not necessary to give your goldfish additional vitamins at all.

If your fish are sick or under a lot of stress, you can also give them daily vitamins. Nevertheless, you should be measured in your approach and not to overdo it. More is not always more, even with fish.

GOLDFISH NEED SPACE

They may be considered quite easy to care for because they only swim in circles and every now and then emit a few air bubbles, which they send to the surface. Sometimes they also curiously look out of the water or through the glass pane of the aquarium. This is exactly the image that many people have in mind when they think of goldfish or want to get one.

But in fact, they require a lot of space. Therefore, the small round goldfish bowls are cruel to animals. Moreover, the animal will not get enough oxygen and the round shape disturbs the orientation of the fish.

Even if it is now clear that you can discard the round bowls, it is still not enough to fill a larger container with water and feed the fish from time to time. In the coming chapters you will get an overview of what further measures are necessary to offer the goldfish a species-appropriate home.

THE GOLDFISH DOES NOT TOLERATE NOISE

It is hard to believe, but the small robust goldfish actually does not tolerate noise. If their environment is quite loud, they can even become hard of hearing. Noise for as little as ten minutes can make the goldfish unable to hear soft sounds. It can take several weeks before the hearing loss subsides. During this time, the fish have severe problems with orientation and mate selection.

A goldfish has exceptionally good hearing, which is why it is so sensitive to loud noises. The swim bladder has the effect of an acoustic amplifier. Sound waves are picked up through it and transmitted to the goldfish's inner ear.

Researchers Michael Smith, Andrew Kane and Arthur Popper investigated the effects that noise can have on fish. To do this, they used an underwater loudspeaker. This provided loud water noise. The researchers found that just ten minutes under these conditions was enough to release stress hormones in the fish and that this short time also ensured that their hearing was impaired.

The stress went away with prolonged listening to the noise, but the hearing damage increased more and more. It took several weeks before they could hear well again.

THE WATER FOR THE AQUARIUM

Twenty to fifty liters per fish, that's how much water the tank should be able to hold. It is best to consult a specialist on the purchase and think in advance very carefully how many fish you want to get. It is important that you buy at least three goldfish, as they would not tolerate loneliness well.

Once you have decided on a large enough aquarium or pond, the next thing you are sure to think about is what water to use.

The answer to the question is simple: you can use ordinary tap water. Ideally, you should fill the tank six weeks before you put the fish in. This allows bacteria to form, which the fish need.

THE FILTER

To keep the water quality constant at all times, it is important that you buy a filter. Here it is important to ensure that this is matched to the size of the pool. A filter has the advantage that the water is cleaned mechanically, but biologically.

Why is this necessary at all? Well, even a fish sometimes has to excrete their waste products, and since it doesn't have a device for doing so like we do, the water becomes contaminated more quickly as a result. But food that is not eaten immediately and floats around in the water can also significantly affect the quality of the water.

Always pay attention to how the filter is working. If it no longer cleans optimally or ensures a good water flow, it should be cleaned, because the constant cleaning of the water causes dirt to stick to it.

To clean it, you can simply swirl it back and forth in the aquarium water and also clean it with a fine brush if necessary. In this way, you ensure that the bacteria remain contained in the water. You can also clean it under running water in the sink if you feel it is not getting clean properly.

REGULAR WATER CHANGE

It is not only important that you regularly clean the filter. Sometimes it is also necessary to change the water. However, here you do not need to change the entire contents of the aquarium. It is enough if you change ten to thirty percent of the water. For this purpose, a three-week rhythm is appropriate.

To clean the panes, you can buy a cleaning magnet in specialized stores, which can use once a week. This is important because the dirt can settle on the discs very quickly. However, it is also the case that the bottom of the aquarium is just as much a place where dirt collects. Therefore, every now and then it is necessary to clean the entire tank.

You will be asking yourself where to put your fish while you clean the tank. In the best case you have a spare aquarium. But you can also put the fish in one or - if you have a lot of fish - in several large bowls. Cleaning does not take too long.

THE POND CLEANING

If you keep your fish in a pond, you also have to lend a hand and clean it from time to time. However, there is one important rule here: always measure the water temperature. This must be at least eleven degrees Celsius. In the autumn and winter months, you should therefore refrain from cleaning. The reason for this is that the fish hold winter torpor. If you start cleaning the pond before March, you would only disturb the animals and cause stress, which is not good for their health.

The cleanup itself involves removing old plant debris such as foliage and dead plant matter. This is of particular importance, because otherwise algae can form and multiply. But also the pond bottom must be cleaned. Here you can get advice from a specialist , who can recommend a good mud vacuum cleaner. So that the pond plants can grow well again, you should also cut them back from time to time.

In spring it is important that you measure the pH value of your pond water regularly - preferably daily. This is optimally between seven and eight. It is equally important to check the carbonate hardness value. This is where the hardness of the water is controlled. The best value is between six and ten °dH (degrees of German hardness). This value is especially important for the development of aquatic life. These tiny animals eat algae and thus prevent them from spreading uncontrollably. A natural filter, if we want to look at it that way.

You can purchase simple test kits for this purpose at garden supply stores. To make sure that your pond is a healthy biotope, it is advisable to check the water values regularly.

Due to snow and rain, the water in the pond is often diluted with soft, low-calcium water during the cold season. The rainwater falls directly into it, the snow lies on the thin layer of ice and both melt when temperatures are too warm. This is then the reason why the carbonate hardness values can be too low in spring. If it is below 5°dH, you need to add mineral substances to it so that the values can stabilize again. You can also obtain these from specialist dealers.

NEW PLANTS

Of course, most often you will want to start straight away in the spring to spruce up the garden and pond, but this is not recommended here, because the fish are not yet active again, and one or two nights could bring frost.

The best time to add new plants to the garden pond is the month of May. Now you can plant water lilies, or sea grass. A trick to keep them from floating up is to put some gravel on their roots. This acts like a weight and makes it possible for the plants to grow well.

Crab claws, frogbit or duckweed can also be integrated into your pond. They are not only a beautiful decoration, but also ensure that the algae cannot multiply.

SELF-SERVICE, THE BUFFET IS OPEN

To prevent your goldfish from completely eating the plants right away, a good mix is recommended.

Plant one side with hardy plants such as eelgrass, which the goldfish do not like to eat. On the other side, you can plant fast growing plants such as waterweed. These are tasty treats for the goldfish.

SAFETY FOR THE POND

A garden pond can be a wonderful eye-catcher. But it can also be a source of danger, especially for small children or other animals. Therefore, it is important that you regularly perform a safety check.

For example, you can plant a few roses around the pond. Children often shy away from this because they fear the thorns. Another option is to use so-called structural steel mesh, which is inserted directly under the water surface or alternatively a wire fence. But, and not only pond owners should adhere to this: As a general rule, children should never play unsupervised near water. Even the best protection can never completely rule out danger.

DO GOLDFISH NEED HEATING?

Goldfish are robust, goldfish are the wild boars among fish - but what about questions like: Do goldfish need heating? The answer is: No. Goldfish feel most comfortable when the water is around sixteen to twenty degrees.

However, if you have your aquarium in a room where it is very cold, you should always make sure that the water does not fall below this very temperature. In that case, a heater is indeed quite appropriate. But let's face it, what's the point of putting an aquarium in a room where it's very cold and where we don't really feel like watching the fish swim?

THE INSTALLATION OF THE AQUARIUM

Setting up the aquarium is practically like thinking about what we might like in our own home. This means that the peculiarities and preferences of the fish should always be taken into account.

Goldfish are lively and love to swim. Therefore, it is important that they also have a sufficient free swimming area available for this. A free bottom area also plays an important role here, because goldfish love to dig.

Therefore, it is recommended to put the following at the bottom of your tank: Sand or gravel, roots, aquatic plants, stones. The green plants provide a certain amount of variety and keep the goldfish busy, as they like to nibble around and like to be able to hide.

THE COMPANIONS OF THE GOLDFISH

An aquarium is not only a piece of nature in your home or a wonderful pastime. It can equally be a colorful, fun fish commune. Goldfish are very sociable, you already know that. But how does it behave with other fish or animals? Can they actually live together with others? Let's take a look together here.

Other goldfish:

If we would like to make a funny movie about the life of a goldfish, the title could be, "One goldfish never comes alone", with the subtitle, "Buy more, will make him very happy."

In fact, you should get several goldfish, because they love the company of others. In addition, a fish in the aquarium can also quickly become bored if it has to swim around alone. This is not very suitable for its species!

The various breeding forms often have differences in terms of their type and intensity of movement, so find out from your specialist exactly which forms can be kept together. The reason for this is that the faster fish can eat away some of the food from the slower ones. A wonderful and above all uncomplicated variation is, for example, the normal goldfish with the veiltail.

Other fish:

In fact, it is better if a pure goldfish tank is created. The reason for this is not that the goldfish do not get along with other species. Rather, it is because goldfish prefer subtropical temperatures. Tropical fish, however, do not tolerate them.

If you think that the domestic cold-water fish can be brought into the fish community, you are wrong, because they do not like the warmer temperatures. At a stretch, goldfish can certainly be placed in a cold water tank, but here the technical effort is very high.

Keepers who still do not want a pure goldfish tank can get the following other fish:

- Rosy barbs
- Gold barbs
- Odessa Barb
- Zebra danios
- Weather loach
- Asian counterparts (it is important that the tank is large enough and has a sand bottom)
- Sunbleak

Catfish are beautiful to look at, but they can be quite dangerous for the goldfish. There is no guarantee that the catfish will not put the beautiful ornamental fish on its menu.

Many of the species mentioned are schooling fish. Therefore, it is important that your tank is large enough for all of them. Imagine that you have to live together with different families in a very small space. You would certainly not like that.

Invertebrates:

Snails are wonderful aquarium animals. Most of them are even excellent to keep together with goldfish. Mass reproduction of snails need not be feared, because the goldfish provide a remedy. Decorative and robust are swamp snails and post horn snails.

You can also try out whether goldfish want to live together with shrimp. However, you should plan several weeks for this and observe the animals well. Unfortunately, it often happens that the goldfish literally eat the shrimp.

Crayfish that are compatible with plants can live together with the popular ornamental fish quite well. If you want to keep them together with comet tails, a sufficiently large tank is necessary. Each needs its own free space. However, if you have compressed, long-finned breeding forms in your tank, you should not keep crayfish.

EVEN A GOLDFISH WANTS TO BE PAMPERED

For sure you know that you can wonderfully spoil your pets like dogs and cats. A treat here, a new toy there, and a tender caress when the animal joins you on the couch. But is it also possible to spoil your goldfish? After all, you can't give them a toy or scratch their bellies.

But, surprisingly, you can actually also spoil your goldfish! You wonder how to do that, apart from feeding them special food? Well, then you should definitely read on.

Goldfish are very curious by nature and will pop up on the surface of your pond from time to time as soon as they get to know you and you stop by the pond. Often the beautiful ornamental fish then cavort on the surface and literally beg you for food, because they have remembered that when someone comes, there is something tasty to eat.

Every now and then, therefore, you can give your fish an extra portion of food. The important thing here is that you stay close by during the feeding time and remove the remains immediately to maintain the cleanliness of the pond.

Another way to pamper the fish is to add new plants, which serve as food, oxygen providers and, above all, as hiding places for the fish. There, they can happily play with the other fish as they love to chase each other through the cool water.

Pampering also includes regularly checking the water values. Adding fresh tap water is also fun for the fish, because they can chase the water bubbles as the water flows into the pool.

And even if you yourself do not see what is happening at the bottom of the pond, you can also put there some great hiding places. The decorative items, which you can use for the aquarium, can also be used for the garden pond.

So, what are you waiting for? Give it a try and provide some variety and entertainment for your fish.

HOW TO BURY A GOLDFISH

Many people may think that flushing dead goldfish down the toilet is not a problem. It may not be, but it is not nice either.

It is the same with goldfish as with any other pet. We become attached to them, even though we can't pet or cuddle them like dogs and cats. And yet it really upsets us when one of our beloved goldfish dies. Anyone who has children should approach the subject with particular sensitivity.

When the beloved pet has died, many think about what to do next. How do you say goodbye to your goldfish with dignity? I would like to tell you something about this:

It is interesting that for many thousands of years not only people have been ritually buried, but also animals. Anyone familiar with the history of ancient Egypt knows that cats and bulls were considered sacred there. For this reason, they were also carefully embalmed before their burial. Another example is the burial of horses in 700 to 800 AD. While these were not buried in the sense that we know a burial, they were carefully set upright. This speaks of true love of animals.

These days, the burial of a pet is not a cultural affair. It offers the owner much more of an opportunity to say goodbye to it emotionally and with dignity. A funeral can also help the pain of the loss to be better processed.

Interestingly, there are numerous ways for pet owners to bury their beloved animal. I would like to introduce you to a few.

The burial

I too was amazed when I read that there are pet cemeteries and even about 120 of them in Germany! Since this book speaks about goldfish, it is not necessary to mention the conditions for burial in the ground on your own property. The goldfish is a small animal and its burial in the garden is therefore not a problem.

Dig a hole, preferably deep enough so that the animal is not accidentally dug up again later. How you perform the burial ceremony is up to you. If you have children, you should involve them and could give the goldfish a dignified burial with music, candles or whatever else you want.

You can wrap the dead little fish in a cloth and put it in a small box. The children can put painted pictures or even letters inside. Close the box. Tip: The box can be decorated just as beautifully.

Then everything can proceed as it is customary. Flowers and sand are placed in the grave, perhaps a few drops of water, because, after all, it is a goldfish, and the grave is closed. In memory, you can put on the grave a small cross or a stone, perhaps a shell or other things that remind of its habitat.

Cremation:

Yes, this is also possible. There are even beautiful urns for this and you can take them home. In addition, there are different variations and services from which to choose.

The diamond burial

This option is a little more expensive. In order for an artificial diamond to be made, the ashes and some fur from the animal is needed. A fish, of course, does not have fur, but many suppliers offer other options for this. The price for the diamond production is about € 1,500, but of course it can vary. You could have the diamond made into a piece of jewelry, for example a ring or pendant, so that you have your pet with you wherever you go.

The burial at sea

In this case it is necessary to have the animal cremated, as this can only be done with an urn. I am also sure that you do not want to put your goldfish just like that into the water, because then it will certainly be eaten by other animals. The costs for this are of course also somewhat higher.

The space burial

Does not exist? I will prove you wrong, because this way of burying an animal does exist. However, the costs for this are at least € 5,000 and few people can afford that.

- Chapter 6 -

PURCHASE AND COSTS

Now that you have intensively studied the animal, its peculiarities, its preferences, its mating behavior and care, as well as many other points, it is time to think about the acquisition and maintenance of the goldfish. These two points are also of great importance before you actually decide to keep goldfish.

PURCHASE AND RUNNING COSTS OF THE AQUARIUM

When it comes to the best place to buy your future fish, there are many options, in the specialized trade, at the breeder, in the pet store or even online. But which is the right way? What should interested parties choose?

First of all, everyone has to figure this out for himself, because the purchase is always associated with costs. The financing of the aquarium or pond itself and its accessories from pump to filter can certainly be enormous. Here, some of you may think that you do not want to spend quite so much for your pets. Understandable, if we look at the total costs, which are described in more detail in the next point.

And yet, here too, it is advisable not to choose the cheapest seller and instead opt for a specialist. A specialist here can be a zoologist or consultant in the pet shop. Both are trained and can have a comprehensive consultation with you.

If you want to buy breeding fish, the best contact here is definitely a breeder. From him you will get all the information you need about breeding fish. In addition, he will have the breeding papers available.

But be careful here too, because not everyone who calls himself a breeder is also a real breeder. Check in advance whether he is in a breeding club. If he tells you the club by name, you can contact them and ask if the breeder is really a member.

Here is a small checklist to help you identify a good and reputable breeder.

1) A reasonable price is quoted.
2) He will allow you to see the important papers.
3) Visiting appointments are made possible.
4) The aquarium or pond is well-maintained.
5) The breeder inquires where the fish are going.
6) A written contract of sale is drawn up.
7) The animals are examined and vaccinated in advance.
8) The breeder is a member of a breeding club.

Entertainment:

Under this heading I would like to show you what you need for an aquarium or pond in which you want to keep goldfish. In addition, I will provide you with a list of approximate costs. Here's a tip: It's just a rough estimate. When you decide to buy goldfish and everything you need, it always matters where you shop. Sometimes we get something cheaper, but sometimes we have to spend a little more.

The fish:

Goldfish are schooling animals. Therefore, you should buy at least three fish. Depending on the size, species and age, the price can range from € 0.50 to € 25.00. However, if it should be breeding fish, you must expect to pay at least € 300 per fish.

The aquarium:

Since you should buy at least three fish, the aquarium must also have the corresponding size. After all, you need 30 liters of water per fish.

Likewise, it is important to pay attention to the quality of the tank. It is optimal if the aquarium already has a filter and a powerful air pump.

A good aquarium, which can hold about 90 liters of water, costs about € 90 to 120.

The aquarium sand:

The sand not only gives support to the plants, but also filters the water in a natural way. For good sand you need to plan about € 12.

The plants:

Plants in the aquarium are not only nice to look at, but also offer the fish additional hiding places. In addition, the fish also like to be able to nibble on something. Therefore, pay attention to variety and also buy some plants that the fish prefer not to eat.

The price for a bunch of about 80 aquatic plants is about € 20.

The water purifier:

With this you can always clean the water from harmful substances. It costs about € 30.

The gravel:

Not only aquarium sand is necessary, but also some gravel. This gives the plants the necessary support and prevents them from sliding out of the ground.

The cost here is about € 15 - 30.

The dip net:

Yes, a dip net is also necessary for an aquarium. For example, if you want to clean the tank completely and need to put the fish in a separate tank during this time or you need to remove food remains. The price for a dip net is about € 5.00.

Decoration:

Of course, we want to offer the fish not only a natural habitat. Such an aquarium should also be an eye-catcher. This is where the decoration comes into play. Everyone has his own taste here. A few examples follow:

- Corals: € 15
- Castle: € 15
- Tree trunk: € 15

The pump and the filter:

Both are very important for an aquarium, because they ensure that the water remains clean and in motion. A filter also reduces the formation of algae. Pump and filter each cost from € 15 upwards. Here, the size and capacity of the tank plays an important role.

Lighting:

Since the fish lack natural sunlight when they live in an aquarium, the lighting is of course also important. Often fluorescent tubes (T8 tubes) are used here, which ensure a day-night rhythm. These are available from as little as € 10.

The gravel cleaner:

This is important to free the soil from debris and harmful bacteria. The cost of a gravel cleaner starts at € 5.00.

The cleaning magnet:

This is also important to clean the aquarium in the best possible way. It is also easy to use, as it is simply attached to both sides of the glass (inside and outside) and you simply move the magnet back and forth as if you were cleaning the outside windows. The price here starts at € 5.00.

COSTS FOR PURCHASING AND CREATING YOUR GARDEN POND

Maybe you do not decide on an aquarium, but on a garden pond. To help you decide, this part of the chapter will show you what the rough costs may be. Again, prices may vary depending on your choice of market. The points shown are only a rough guide.

It is important in advance to think carefully whether you want to create the pond yourself or hire someone. The cost of the second option will be significantly higher.

The fish:
Goldfish are schooling animals. Therefore, you should buy at least three fish. For a pond you can choose goldfish that are larger, because here, depending on what you decide, the available area is much larger. For this, the price per fish is then also higher than for aquarium fish.

The pond basin:
First, you need to decide on the type of pond basin you want. You can choose a ready-made basin, or try making it yourself. However, if you choose a ready-made basin, the cost will be much lower.

The following points should be considered for your decision:

- Pond size

- Pond depth

Five fish can be kept per 150 liters of water. A ready-made tank for this capacity costs about € 40 - 50 per 100 liters of water.

The excavation:

If you have the excavation done by a company through, you can expect the following costs here: Foil pond or prefabricated pond per m^2 € 40 - 130. If you do it yourself, you will of course pay less, but you should plan a lot of time and you need to think about a solution for removal of the enormous amounts of soil.

The sand:

Sand should not be missing in a pond either, as fish love to burrow. Here it also depends on the size of the pond, because a larger basin needs more sand.

For this you have to expect costs from € 15.

The foils, the fleece:

The pond liner is the protective shield between the water and the substrate. If you are building the pond yourself, the foil forms the basis for a watertight pond. The cost of this is € 27 - 53 per m^2, depending on the size.

The pond liner must also be protected. Therefore, it is important that you first lay out sufficient fleece under it. The cost of this starts at € 25 per square meter.

The plants:

Plants in the pond are not only beautiful to look at, but also provide additional hiding places for the fish. The price for the aquatic plants is about € 20 for a bundle of 80 plants, as with the aquarium.

Furthermore, you can also add various plants around your pond. This makes it all look even more natural. It creates a feel-good oasis that is wonderful not only for fish, but also for yourself. Depending on what you want to plant and how much space you have available for it, prices will vary. Additionally, it depends on where you want to buy the plants. It is possible to buy good plants for about € 50.

The gravel:

Gravel is also important for a garden pond. As already mentioned with the aquarium, this gives the plants the necessary support and prevents them from sliding out of the ground. The cost here is about € 15 - 30.

In addition, you can also spread gravel around the pond as decoration. Since a larger amount is needed here, the costs for this will be a little higher.

Decoration:

In a pond we do not need decoration, because we cannot see it anyway, and it is already quite similar to the natural habitat of the goldfish.

Here you can add your own designs to the appearance of the pond with a few nice decorations and some ambient lighting. There is a lot of choice and how much you add is entirely up to you. The cost for a simple pond will start from between € 50 and € 100.

The pump and the filter:

It may sound strange, since the natural habitat of fish does without filters and pumps, but for a pond in which fish live, a filtration system is really necessary. The reason for this is that every now and then there will be food remains floating around in the pond.

You have to look at it this way: A garden pond is a confined water area and the fish population lives closer together here than in a natural body of water. Therefore, nature cannot manage to maintain the quality of the water on its own.

You can buy a simple flow-through filter for as little as € 40. It is best to seek advice from a specialist store, as the size of the pond must also be taken into account here.

The gravel cleaner:

Here, too, it is important to support nature, as the garden pond is a self-contained body of water. Therefore, you also need a gravel cleaner for a garden pond.

One of these will cost about € 50, for a simple one, although they can be much more expensive.

The dip net:

Surely leaves can sometimes land in your garden and fall into the water too. Sometimes small animals also get lost in the water or drown when they try to drink from the pond. To remove them, a dip net is a good choice. The price is about € 5.00.

GARDEN POND OR AQUARIUM

Now you have received some information about the aquarium and also about the garden pond. You also have an idea of the costs involved in purchasing your goldfish, but this is by no means the end of the story. There are other costs, such as electricity and fresh water when we want to change it. The amount you spend varies greatly, depending on all the factors mentioned below.

- Residential area
- Size
- Filling quantity
- Electricity supplier and tariff
- Quality of the pump and filter
- Number of water changes
- Water quantity
- Energy class of electrical devices

All these factors must be taken into account. It is therefore advisable to note down all the costs and create a rough comparison. Take your time, as the purchase is only the beginning and does not include any work and costs which the future may bring. You should be clear about what you want and if you want to go ahead with those plans.

How do you want to keep your fish? Will you choose the cheaper variation or the one that most closely resembles the natural habitat of the fish? Some goldfish keepers choose a combination of both. In the warm seasons the fish live in the garden pond. When these end, their keepers transfer them to an aquarium where they can spend the winter. However, this variation is not the best, because fish also have an internal clock, which tells them that it is time to rest. Therefore, if you do not have a possibility to let the fish hibernate in a cool place (that means below ten degrees Celsius) and in peace, it is better not to choose this combination.

Especially in our present environment, the temperatures in the winter months are no longer so low that a fish could not survive in the garden pond. In addition, the relatively mild winter temperatures also make it difficult to even find the right time to make the change.

In our houses, the temperatures are also very warm, which would affect the natural rhythm of the fish. In this case, the fish then become more active and their metabolism increases, at a time when it should be ramping down. This often leads to the fish suddenly displaying mating behavior.

So if we were to ask a goldfish and it could answer itself, it would certainly choose a pond, as it is the ideal substitute habitat. Most fish also reproduce only in ponds without much help from the keeper.

And yet, from a human point of view, it is often not recommended to keep goldfish in a pond. They are quite small and we ourselves do not really see much of them. For this reason, many people prefer ornamental carp, the so-called koi. These are larger and we can see them more often. For this reason, it is more often recommended to keep your goldfish in an aquarium.

As you can see, it is quite difficult to make a general recommendation here. In the end, everyone has to decide for himself. In order to support you in your decision, I would like to present both variations in methods in more detail.

Keeping them in the pond:

Those interested in goldfish will already know that keeping them in a pond is more suitable for the elongated species. Goldfish with round bodies tend to prefer warmer water, and they are also more susceptible to disease. Their physical characteristics are also not conducive to independently picking up food in the pond and moving around appropriately. Simple goldfish can be kept in a pond without much difficulty if the following important rules are followed:

- Not too many fish at once
- Adequate feeding (small pond means there is less food available and humans have to feed in addition)

- It needs a good filter
- The pond should have a minimum depth of one meter
- In winter, a small area of the ice layer should be left open to allow oxygen supply and the escape of fermentation gases.
- In fall, make sure that fallen leaves are removed quickly
- Use plants which remove nutrients and metabolic substances from the water.

Keeping them in the aquarium:

Those who want to keep their fish in an aquarium do so for a good reason: they want to see them and dream about them. And yet it is very critical and difficult to recreate the natural habitat of fish in a tank, which is much smaller.

To ensure a good life for the fish with plenty of variety, you should study and take into account their basic needs:

- Sufficient space
- Various plants
- Sand and gravel bottom for burrowing
- At least three fish

Goldfish can become very old, so you should think carefully whether you really want to keep goldfish. For this reason, it is advisable not to buy them spontaneously and certainly not as a gift.

- Chapter 7 -

DISEASES

Even a goldfish can sometimes get sick. In this chapter, you will learn what diseases it can get, how it gets them and how to treat your goldfish. In addition, I would like to introduce you to some disease prevention measures.

Before we get to what diseases goldfish can get at all, let's first deal with the subject of oxygen: fish get along quite well if the water does not contain too much oxygen. But - and you should remember this, because it can also happen to you - it can clearly be seen when there is too little oxygen. How? They swim to the water surface and gasp for air.

This is where your action is needed. But before we get to that, we should first look at how oxygen deficiency occurs in the first place.

BREATHING

Of course, goldfish also need oxygen. Without it, they could not survive either. However, fish do not have a nose like we humans or other animals. Their respiratory devices are the gills. When a fish opens its mouth underwater, the water flows in and with it the oxygen. The gills filter out the oxygen and carry the water back out. Carbon dioxide is also expelled with the water.

What causes the gasping for air?
In order to be supplied with oxygen, the fish must constantly have water flowing through its gills. So if the water in your pond is too low in oxygen or the fish's gills are not functioning properly, it is at risk of dying an agonizing death by suffocation.

So if it swims to the surface and gasps for air, this is a sure sign that the oxygen level is too low. The fish then gasps for air at the water surface, the gills suck up the oxygen and expel the remaining air in the form of small bubbles.

Reasons for too little oxygen in the water:

1) When temperatures rise, the oxygen content in the water decreases.
2) Too few plants have been planted. Therefore, no new oxygen is produced.

3) There are too many fish swimming in the pond.
4) There is too much food or excrement at the bottom of the pond, which can cause rotting.
5) The filter is not working properly due to rotting or insufficient water flow.

Sometimes there are only a few fish gasping for air at the surface. In this case, the oxygen content is not the problem. Problems with the gills could be the cause.

These could be, for example:

- Parasites
- Gill worms
- Burns due to low pH value

FIRST AID MEASURES IN CASE OF OXYGEN DEFICIENCY

If you observe that your goldfish are gasping for air, this is a sign that your help is urgently needed and that something is wrong. Usually this indicates that the oxygen level in the water is too low. Here it is advisable to put in a membrane pump. Those who already have an aquarium will be well acquainted with such a pump. This is connected to a hose and a so-called outflow pond. The stone ensures that fine bubbly air flows into the water. In addition, direct fresh water from the tap into your pond, because this also contains new oxygen.

You can also add oxygen tablets to the water. For now, all the steps just mentioned should be sufficient to increase the oxygen content of the water so that your fish no longer have to gasp for oxygen at the upper surface. However, continue to monitor your fish, as other causes may also be playing a role in this behavior.

FURTHER MEASURES

To ensure that the oxygen content of your pond water does not drop again after a short time, you can place an oxidizer in the shallow area of the pond. This device contains a cartridge with hydrogen peroxide. This decomposes into water and oxygen, which means that your fish are not exposed to toxic substances. The device itself works without electricity. Another advantage is that the device comes to the surface by itself as soon as the cartridge is empty.

It is best to consult a specialist for this.

CAUSES OF DISEASES

There is a reason for everything, including goldfish diseases. It is important to understand the cause for the diseases so that you can learn how to prevent them.

One of the most common reasons is that pond owners and goldfish keepers create the wrong or insufficiently good conditions. For example, the water values may not be good or the diet may be wrong.

Stress can also be a trigger for diseases in fish. This can happen, for example, if the pond is too small or there is too much stock. Other causes could be parasites or germs, which could occur through incorrect pH values of the water.

Sometimes we also buy a new fish. But here caution is required! It could carry pathogens, which could then be passed on to the other fish. Therefore, it is advisable to put the new additions in a quarantine tank first. There we can observe them well and see if they show any abnormalities.

Tip: If you observe your fish frequently, you will train your eye to recognize diseases or abnormalities early and be able to act in time.

SYMPTOMS

Detecting diseases is not too difficult even in fish. Basically, you just need to know what to look for. So that you too can always ensure the health of your fish, we will now take a closer look at the symptoms:

- Lingering on the bottom or on the surface for a long time
- Swimming disorder
- Headstand
- Scouring at the bottom
- Swim surges
- Fish jumping when itching
- Rapid movement of the gills
- Respiratory distress – breathing at the surface
- Physical changes (color, coating on the scales)
- Weight loss
- Changed body shape

Important: If you notice many symptoms, rapid action is required, as pathogens multiply more quickly in the water.

COMMON DISEASES IN GOLDFISH

Parasites / White spot disease (Ichthyophthirius multifiliis):

Ichthyo is a parasite that causes white spots on the mucous membrane of the goldfish. This is where the name white spot disease comes from. However, it can also happen that the eyes of the fish are affected. This disease causes respiratory distress in fish.

The parasite can multiply very rapidly. Pockmarked growths can be seen on the body of the fish, which fall off over time and sink to the bottom. It is not hard to imagine that these parasites would grow significantly in numbers and could easily be passed on to the rest of the fish. In 24 hours, it can reproduce up to a 1000 times in the water.

With this disease, it is therefore enormously important to start treatment as soon as possible.

You can give the fish malachite green to eat for five days. This usually helps to get rid of the parasites quickly. However, sometimes it may be necessary to continue the treatment longer.

Costia (called Ichthyobodo Necator in technical terminology):

This parasite causes the skin of the fish to become cloudy. They are the classic weakening parasites. However, these parasites only have an easy time with young or weakened fish. Adult fish are only attacked by them when their immune system is weakened.

The infestation of fish by these parasites can also be favored by a water temperature below 15 degrees Celsius. Pond owners should therefore always keep an eye on the water temperature. If the temperature has dropped below 15 degrees Celsius, these single-celled organisms begin to irritate the mucous membranes of the fish. This can be recognized by a white-bluish coloration on the fish skin.

But the parasite infestation is not the only problem in this case. The irritation of the mucous membranes makes the fish more susceptible to fungal growth. Unfortunately, a high infestation then often leads to the fish dying.

For treatment, sometimes simple salt baths are enough. Through these, the metabolism of the fish is supported, which in turn leads to their immune system becoming strengthened.

Before treating the fish with medication, the causes of the parasite infestation should first be clarified. Only if we know the reason for the Costia infestation, can we eliminate it permanently and finally get rid of it. If you are unsure about this, it is advisable to contact a veterinarian who specializes in fish.

Flukes (Gyrodactylus spp., Dactylogyrus spp.):

They are tiny and yet they can be very annoying for your pond inhabitants. As a rule, we humans cannot see flukes with the naked eye. In fact, they can only be reliably detected under a microscope.

Flukes are divided into two types:

- Skin fluke (Gyrodactylus spp.)
- Gill fluke (Dactylogyrus spp.)

The former is found on the outer skin of the fish. Since it damages the mucous membrane of the fish, the fish can suffer other problems from it, such as inflammation of the scales. In addition, the goldfish are more quickly affected by fungal or algal infestation.

A sign that the fish is indeed suffering from the skin fluke is if it keeps chafing and even starts jumping occasionally. You can also recognize an infestation of skin flukes as the skin appears cloudy.

The gill fluke, as its name suggests, is found on the gills of the fish. Fish that suffer from this worm very often have difficulty breathing, as the gills are responsible for this and become inflamed and swollen due to the worm.

To combat the flukes, helpful remedies can be used. If the infestation is very small, simple salt baths are also sufficient to get rid of the flukes. However, if the infestation gets out of hand, it is safer to consult a veterinarian who specializes in fish in order to be able to purchase the right preparations for treatment after a reliable diagnosis. Usually he recommends buying the preparations in a specialized store or he prescribes medication directly, if the infestation is very high and the fish are suffering enormously.

You should absolutely avoid treatments with antiparasitic agents without consulting a doctor, because excessive use causes increased resistance!

Carp louse:

The name suggests that this is a type of louse, but in fact, a carp louse is a crustacean. They grow to about 13 mm in size and can be easily seen with the naked eye.

If the fish did not have to suffer from them, their entry into the domestic garden pond would be funny , because they are brought by waterfowl.

Once they have entered the water, they attach themselves to the scales and the skin of the fish. Once they have done so, the skin becomes reddened, as they cause inflammation in the fish. It is also the case that the fish experience massive itching. If you notice that your fish constantly chafe or shoot uncontrollably through the water, you can assume that the carp louse is responsible.

If only individual lice have entered the pond, they can be collected quite easily. However, as soon as the infestation is larger, it is important to treat the fish with medication. Experienced fish keepers can use over-the-counter remedies, which they can find in specialized shops. For beginners, it is advisable to first consult a veterinarian who specializes in fish. He will then prescribe a remedy.

Fungi / fish mold (Saprolegnia parasitica):

A ubiquitous pathogen is the fish mold. If your fish have a healthy immune system, it cannot harm them. It is different and even very dicey if your fish are susceptible to diseases. Another reason can be the low temperature in winter. This also ensures that the immune system of the fish is weaker than usual. If fish have wounds that exist for a longer time, they are also very susceptible to fish mold. This often overgrows the wounds. You can recognize this by the fact that the wounds are covered with cottonwool-like coatings. Initially, these have a whitish color. However, it is possible that they take on a grey-greenish color. Unfortunately, it is very common for fish spawn to be affected, resulting in loss. You can apply higher concentrated short-term salt baths or buy remedies containing malachite green from a professional.

Bacterial diseases:

The course of a bacterial disease as well as its appearance is very multifaceted. Bacterial diseases are often caused by bacteria that live in the pond anyway. These usually do not cause any harm. However, when the water quality deteriorates, this works in favor of the bacteria and they begin to multiply, which in turn causes diseases in the fish. Depending on the symptoms which appear, the disease could be one of the following:

Hole-in-the-head disease:

This disease is mainly caused by bacteria. In addition, it can also occur due to other pathogens or environmental conditions that are not right. We recognize the disease when the fish show large holes, which resemble ulcers, on the skin. Usually they can be seen on the trunk or on the tail fin. In the worst case we can see the muscles of the fish.

How quickly the disease progresses depends on the trigger and environmental conditions. It is possible that you as a pond owner can expect high losses and sudden deaths of the fish. If this happens, you should consult a specialized veterinarian as soon as possible. The veterinarian will take a swab to determine the cause of the disease. In addition, he will then carry out a resistance test and initiate a therapy.

Fin rot:

The symptoms of this disease are very diverse. On the one hand, the fish may have milky or red discolorations on the edges of the fins. On the other hand, these can also look frayed. If this disease occurs, the keeping conditions are not right. But it can also be that only single fish are affected. Here the reason can also be a local injury. Also with this disease it is advisable to seek help from a veterinarian. After all, it can never be excluded that other pathogens also play a role. A diagnosis is necessary before treating the fish. Only then can the correct medication be selected.

Viral diseases:

Koi herpes virus (KHV):

This disease has only been described for about twenty years. Since it is an epidemic disease, keepers are obliged to report it to the competent authorities.

The most noticeable symptoms are mainly the severe damage that the affected fish has on its gills. However, it is also the case that the virus can affect other organs such as the skin, intestines or kidneys. Unfortunately, sometimes we do not even see the infection.

Temperatures between 16-28 °C, but also stress, which the fish suffers, can cause the outbreak of the disease. Fish that are infected often show loss of appetite, visible damage to the skin and they are also more susceptible to other diseases.

One noticeable symptom of this disease is that they suffer from shortness of breath, due to the damage caused to their gills. You can recognize their distress when they rise to the surface or swim close to the water filter outlet gasping for air. Unfortunately, many fish will not survive this disease. There is no treatment available to deal with the root cause of this virus; their only hope is that conditions within the pond will improve. A veterinarian will take tissue samples directly from the gills to determine the disease.

Carp Pox/ Koi Pox (CHV-1):
As soon as the temperatures in the pond drop, this disease can sometimes be observed. The affected fish have a whitish, waxy buildup on their skin. This can also occur on the fins of the fish.

These growths grow in fish as soon as their immune system is weakened. Poor weather conditions and cold water are the main causes here.

Since the fish carry the virus permanently in their bodies, a complete cure is not possible. We can only ensure that the growths recede and create the best possible living conditions for the fish to reduce renewed outbreaks.

Other diseases:

Swim bladder inflammation:
If the swim bladder is inflamed, the fish can no longer provide buoyancy on its own. In the worst case, it will lie on the bottom of the pond. The fish must now rely on the power of its fins to move at all, and if so, only with difficulty.

For this, the fish has to exert an enormous amount of force and gets calluses from lying down. This disease usually affects only a few fish.

The causes can be metabolic disorders, parasites or even bacteria. The type of treatment chosen depends on what caused the condition. This is again determined by a specialized veterinarian. You can also increase the temperature of the water and add iodine-free table salt to the water, which supports the kidney function of the fish. Unfortunately, a fish which has already fallen ill with it, can also fall ill again.

Energy Deficiency Syndrome (EDS):

This disease is one that occurs quite classically in the spring. Here, a deficit between required and available energy is the cause.

Reasons for energy deficiency:

- Poor nutrition (summer)
- Early cessation of feeding to winter
- Low oxygen concentration

Fish that are emaciated are therefore among those most likely to be at risk. You may find it hard to believe, but EDS can also occur in fish that are too fat. The reason is that they can hardly use their fat reserves during the cold months.

Fish suffering from EDS usually swim uncoordinated and rather slowly. They also have shallow breathing. In addition, their kidney function is restricted. As a result, a lot of water accumulates in their bodies, making them bloated and puffy. The scales of the fish stand up like pine needles.

To help the affected fish, the water is slowly warmed salt is added. This stimulates the kidney function of the fish. It is also advisable not to give easily digestible food until the fish's behavior returns to normal. EDS is an emergency and should not be downplayed at all. Unfortunately, deaths cannot be ruled out in this case either.

HOW TO TREAT GOLDFISH

When fish are sick, we naturally wonder if we should go directly to a vet. Depending on the illness, we may sometimes be able to treat the fish ourselves. Salt baths, for example, are a good way to do this. But you can also ask in a specialist store for over-the-counter remedies. However, it is often the case that simply changing the water (even halfway) is helpful.

However, if you want or need an accurate diagnosis, you should not rely solely on your knowledge and skills, but seek the help of a professional who can further help you and especially your fish. Please also remember that there is no panacea for all fish diseases. If you treat your fish unnecessarily with medications, you only put additional strain on their organism. This in turn can lead to resistance.

It is also important that you do not consult just any veterinarian, but actually consider one who specializes in fish. He will know what is wrong with your fish and which treatment method is the most suitable.

DISEASE PREVENTION MEASURES

The rule is: prevention is better than cure! You should remember this in order to protect your fish in the best possible way. Diseases affecting fish can multiply quickly in a pond or aquarium and have a fatal effect on the fish population. It is therefore essential to provide the fish with an optimal habitat and to always ensure excellent conditions. The water quality is the most important thing, because the fish cannot simply withdraw from the pond. Good filtration of the water also plays a role here. If you want to keep many fish, you need a large pond. You should also design it appropriately for the selected species and, above all, with plenty of variety.

And not only the correct feeding is important, but also the storage of feed. Therefore, look at the descriptions and instructions on the packaging. But it is always important that the feed is kept dry, cool and protected from air. Thus, this can be protected from spoilage.

If you love your fish, you should think about which vet you can go to in case of an emergency even before you get them, because not every vet actually specializes in fish.

You should never give your fish any medicines to prevent possible diseases. This does not help, on the contrary. The organism of the fish will only be negatively affected. In addition, Pathogens become resistant to the agents which are supposed to cure them.

However, it really does make sense to have the fish examined as a precaution. Find out from your vet whether he offers so-called spring and fall check-ups. You can then have these done without hesitation, because it offers you and your fish security and shows you what you should improve if necessary.

You can check the water values yourself. For this purpose, you can make a list to record how the values stand and when you last checked them. In addition, this enables you to counteract negative changes early enough. Also make sure that the pond / aquarium is well ventilated and the filter system is working properly. This way you guarantee that the water provides enough oxygen for your fish.

When the temperatures outside become lower, they also become lower in the water. During this time, you should refrain from making any changes to the pond unless they are absolutely necessary. At these times, you should also avoid introducing new fish.

Tip: Observe your fish again and again. This gives you the opportunity to learn about their behavior and notice symptoms that indicate a disease. This way you can always take good precautions.

IMPROVEMENT OF THE POND'S MICROCLIMATE

The measures already described are not permanent problem solvers, but only first aid measures. Therefore, it is important to think about how the pond's micro-climate can be improved in the future. In the following text you will therefore find further measures that can help you and especially your fish.

1) Stock

A fish of average size needs about 30-100 liters of water. When creating your pond, you should therefore think carefully about how many fish you want to keep and what size your pond must therefore be at least. It also plays a role whether you might want to add more fish in the future. Precise planning is therefore essential.

2) Shade

Summer is coming and with it the high temperatures. For fish, these can also be dangerous. Therefore, it is important that the pond is in a place where about half of it is in the shade during the midday hours.

But you can also take the following measures, if you have not considered this when creating the pond:

- plant tall grasses
- Insert water lilies and other floating plants

- Plant conifers on the southern bank

3) Aquatic plants as suppliers of oxygen

The pond should be stocked with delicate, feathery aquatic plants at a depth of approx. 50 centimeters. They produce plenty of oxygen and also promote the clarity of the pool. Suitable plants: Hornwort, water milfoil and waterweed.

IS DEWORMING ALSO NECESSARY FOR GOLDFISH?

In the past, the Spanish conquerors brought European epidemics to South America, which often ended fatally. Smallpox or measles even depopulated entire regions. This may be a stretch, but the introduction of diseases by new goldfish can also be compared to this.

If unknown pathogens encounter an otherwise intact immune system, this can have very fatal consequences. Therefore, it is important that every keeper gives his fish the opportunity to get used to the pathogen strains. It is important that the new fish are first quarantined. This way you can see if any fish are sick and sick fish can take their time to heal. If the fish can then be considered healthy, they are added to the old stock. It is important to exchange the water so that all can get used to each other. At the beginning it is useful to take only a syringe or a shot glass of fresh water to see if any changes become visible in the fish. Continue this for a few days. After a week, you can start exchanging larger amounts of water. Consequently, the quarantine does not simply end, but could take a few weeks to complete. A few days after that, you can look at making a more extensive water exchange.

Long-time goldfish lovers would all agree that goldfish should be de-wormed, because they know that a goldfish can contract gill flukes, which can be very uncomfortable for the fish. Take care not to treat your fish with regularity. Too much medication is not good for your fish and could even lead to liver damage. It is also the case that parasites eventually become resistant to regularly administered medication. It can be assumed that gill flukes in particular are always present in small quantities. They only take the upper hand when weakening occurs.

Before you use medication, you should consult a veterinarian, who will examine your fish and, after diagnosing the ailment, can decide what is best for it.

- Chapter 8 -

GOLDFISH BREEDING

If you want to breed goldfish, you should be aware that it is not a piece of cake. You cannot simply select a pair and expect that these fish will be willing to mate.

The pond should be stimulating and provide the fish with a place where they feel safe. It should also provide enough protection for the future young, otherwise the adult fish will eat their spawn shortly after it is laid. However, the adult fish will not only eat the spawn, but also the young.

The steps just described are indispensable for successful breeding, even though they will take a lot of time and cost money. In addition, it takes some patience and practice to actually be able to make profits later on. But this should not deter you at all. Believe in yourself and in your plans, then everything will go well.

CORRECT CONDITIONS

To successfully breed goldfish, you should plan the procedure well in advance. It will take some time for you to create an optimal environment for the future parents and their fry.

The fish should be purchased at least one year before breeding. This will give them enough time to get used to each other and the new environment. In addition, the fish will gratefully accept that they can explore the pond well in all corners and make it their new home.

It is recommended to purchase the goldfish in the summer months of July or August. Then in the spring they are already willing to mate. If you have bought your breeding fish, it is recommended to clean them first. It is best to ask the dealer directly how best to do this.

THE RIGHT ENVIRONMENT

If you decide to breed in an aquarium, you should set it up as naturally as possible. For breeding it is advisable to choose a tank that can hold at least 90 liters.

Natural breeding requires a lush environment. That is, you need to plant a lot of water plants, because that is where the hatchlings hide from the adult fish and take their first food after hatching.

There is also the possibility of breeding fish artificially. This does not require plants. But and you should always keep this in mind: Plants improve their quality of life. If you decide to go the artificial way, you should invest your money in a proper spawning mop. The female fish can then lay their eggs in this. In addition, the eggs are protected there from their adult counterparts, which would otherwise eat them.

FEEDING

If you want to start breeding even though it is not yet spring, you can offer your fish food such as brine shrimp or even live black worms. With this food you can make the fish believe that it is already spring, as this is more abundant in the natural environment in the springtime. In addition, you should make sure to put your fish on a light diet, as they eat little at first in spring due to their winter torpor.

Tips for proper feeding:

- Small but frequent meals
- 3 meals a day
- Do not overfeed
- Remove remains of the feed
- Feed should be crushed

MIMIC TEMPERATURES

In the spring, temperatures tend to be lower than in the summer months. Therefore, if you want to start breeding in a summer month, you should lower the water temperature and increase it again at intervals.

The reason is that goldfish prefer to mate in the spring. This is a good way to deceive them.

Lower temperatures properly:

- 10 - 12 degrees Celsius
- if willing to mate, increase by 2 degrees Celsius per day up to 20 degrees Celsius

DAILY MOVEMENT OF THE WATER

Yes, this is also important for the fish to feel comfortable and ready to mate. A partial water change can be a good support here. In addition, this also has a stimulating effect on the fish. It is therefore recommended to change approx. 20 % of the water daily.

THE GENDER DETERMINATION

It is important that you know what a female gold fish looks like. The only really important task, which is assigned to you if you want to breed fish, is to determine the gender of the fish.

How to recognize the females:

Look at the genital opening of the fish. If you can see an opening between the anus and the anal fin, you have already won half the battle. The female excretes the eggs from this, the male the sperm.

But how can we tell if it is a female or male fish? Quite simple: the genital opening of the female is rounded and curved. In addition, you can also feel the abdomen of the fish. However, you should be careful here so as not to harm it. If the abdomen feels soft, this is a sure sign that the fish is female.

You can also look closely at the pectoral fins. These are short and round in the females. In addition, it is worth mentioning that the females are narrower and rounder than their male counterparts.

How to recognize the males:

A male goldfish is usually smaller than the female. Male goldfish have small white dots on their fins, head and gills when they are ready to mate.

The genital opening of the male is curved inward. If you palpate the abdomen, you will notice that it is stiffer and firmer than the females. The male goldfish also have more pointed and longer pectoral fins than the females.

THE BEHAVIOR

Males behave differently than females. Yes, we know this example from the human world as well. As soon as the mating season arrives for goldfish, you experience - as you already know - a real spectacle, I could also say a drive hunt, because the males literally chase the females through the water. At first everything starts quite harmlessly and calm, but with time the fish become more intense in their love play.

Place a female in the tank and see how the other fish behave. The male fish will show increased interest, while the females will continue to make their rounds in the water rather disinterestedly. Experienced breeders isolate the female fish from the male fish about two weeks before mating season. The reason for this is that it increases the willingness to mate on both sides. What you can't have, you desire, just like us humans.

BREEDING WITH YOUR OWN GOLDFISH

If you want to try breeding with your own fish, choose the best pair among your fish. Make sure that you choose young and robust fish. These are very fertile and more likely to reproduce.

Pay attention to the back and chest area of the female. If this is large, it is an ideal female for breeding. When choosing a male, your choice should be a large male. In addition, it should be a good swimmer, as the act of breeding is like a hunt through the water. Males that have many white spots on their gills are also a good choice.

The ideal is to select the three best males and the two best females for breeding. Once you have made this choice, it is time for the fish to be isolated to have as little distraction or disturbance as possible.

THE ISOLATION

Once the five selected fish are placed in the separate tank, you will notice that the males become lighter colored in the abdominal area. In addition, they swim quickly through the tank and they appear to be quite excited. Then the hunt for the females begins.

In doing so, they repeatedly bump against the body of the females until they eject their eggs above the plants. Immediately afterwards, these are fertilized by the male.

ARTIFICIAL INSEMINATION

Sometimes it also happens that the natural way was not successful. This is where you come in. Place a male and a female fish in a shallow pool. Hold the male fish and then rub his genital opening. This is how you rid him of his sperm. Swirl your hand through the water at so that the sperm is distributed.

Next, do the same with the female. Rub its genital opening and free it from its eggs. Now swirl the water again so that eggs and sperm find each other.

It is important that you are always very careful when doing this, as the fish can be injured very quickly and easily. Apply very little pressure, so you do not cause pain to the fish.

With artificial insemination, it is not necessary to keep the fish under water. They are quite capable of – although not very good at – breathing outside the water. Nevertheless, the rule is not to keep the fish above water for more than 30 seconds. So it is necessary to put them in the water again in between.

SEPARATING THE FISH FROM THE EGGS

Yes, this step is actually necessary if you want to protect the spawn. Goldfish are cannibals and will eat their own spawn just a few seconds later. Therefore, to achieve the best possible breeding result, you should remove the adults from the tank with the eggs.

The young hatch after only a few days. At first they feed on the yolk sack that envelops them. However, they soon begin to nibble on plants in their environment and eat small organisms from the water. You can also offer them the same food you give to the adult fish just a few days after hatching. Here, however, it is important that you chop it well, since their mouths - like everything else - are still very small and fine.

If possible, the eggs should remain in the same water in which they were fertilized. The reason for this is that they are very fragile.

A GREAT NURSERY

Since goldfish hatchlings can lay up to 3000 eggs, it is important that the nursery is of sufficient size. As soon as the young have hatched, they want to move around. And if there is little space, this can significantly impair their natural urge to move and thus their development.

SUMMARY AND INTERBREEDING WITH OTHER FISH SPECIES

In chapter four you have already learned how reproduction works in goldfish. As a reminder, I will briefly summarize this once again:

- April to May is the spawning season.
- Males chase females through the water.
- Females are nudged by males again and again, it resembles a hunt.
- When nudged, egg laying is stimulated.
- The male carries the female to the surface. The spawn is deposited there and immediately fertilized.
- 500 to 3,000 eggs are possible.
- 5 to 7 days later the larvae hatch. They are almost transparent and sticky.
- The larvae attach themselves to the water plants.
- Important: The spawn should be kept safe, otherwise the adult fish will eat the eggs.
- Goldfish only get their color after a year.

In this chapter you have learned a lot about the conditions that are necessary for successful breeding. But what about other fish species? Do goldfish actually mate with other fish species?

Nature has a plan for everything. Different fish species are not meant to reproduce with each other. However, if man intervenes in reproduction and breeding, many things are possible.

At this point I would like to emphasize that you should not experiment. Fish should reproduce in a natural way if possible. In addition, only experts who are actually extremely well versed in breeding fish should dare to make such experiments.

- Chapter 9-

WINTERIZING

In winter, temperatures can drop very low. Many pond owners who are just starting to keep goldfish are then faced with the question of how best to get their goldfish through the winter. In order to provide you with enough information about this as well, this chapter will cover how to winterize your fish. The following text will tell you in five steps what to consider and how best to proceed without causing unnecessary stress to your fish.

A SUITABLE PLACE

Fish owners who have a very large and deep pond can safely let their fish stay in it during the winter, because it does not freeze to the bottom. These beautiful ornamental fish then drop to the bottom and fall into so-called winter torpor.

However, it is important that the pond is actually suitable for winterizing and meets important criteria. If this is not the case, you should set up a suitable place for your fish within your four walls.

Important factors for winterizing the pond:

- The depth of the pond should be more than one meter.
- There must be enough space for all fish, even in the deep places.
- An adequate oxygen supply must be ensured.

WINTER QUARTERS IN THE HOUSE

This method is necessary if your pond does not have sufficient depth. Some goldfish are very sensitive to cold and therefore feel more comfortable in winter quarters in your home. The veiltail and cardinalfish, for example, are particularly sensitive to the cold.

If your fish have lived in your pond for less than half a year, it is also advisable to bring them indoors for the winter. The reason for this is that the fish are not yet accustomed to the new environment. When winter comes, they will not be able to cope as well as the fish that have been in your pond for a longer period of time.

Establishing the winter quarters

If you have decided on winter quarters, it does not mean that you are finished. Now it's time to prepare the aquarium.

However, if you decide to let the stay in the pond over winter, here is a brief guide on how to winterize your pond:

- Clear the pond of leaves, preferably before the first frost in late autumn.
- Some bottom mud may remain, it serves as a retreat.

- Stop feeding when the temperature is around eight degrees Celsius.

If you decide to keep the fish indoors during the winter months, you should bring them in before they go into winter torpor. The best time to do this is before the temperatures drop below ten degrees Celsius.

However, an aquarium that is exposed to warm room temperatures inside your home may not be suitable for the fish. It is better to place the aquarium in a rather cool place. Alternatively, you can simply use a rain barrel A basement or garage is the optimal location for the rain barrel or aquarium. If the temperature of the water is around four to eight degrees Celsius, your fish should get through the winter just fine. Equip the rain barrel additionally with a filter and a pump, so that the fish are also supplied with oxygen in winter.

Of course, such a move to the winter quarters also means stress for the fish. Therefore, two-thirds of the water in the barrel or aquarium should be pond water.

Keep an eye on the winter quarters

For the fish, oxygen is the most important thing for their survival. It is therefore important that you never completely cover the rain barrel, if you decide to use one. This way oxygen can always get in.

Correct feeding

If you take your fish indoors in winter or the temperatures are cool but still mild while your fish spend their winter in the pond, it is appropriate to feed only in small quantities. For this, you should use so-called sink pellets, as the fish spend their time at the bottom of the pond. In addition, they are rich in high-quality fats and have few carbohydrates. The omega fatty acids it contains give the fish strength in winter. Normally, however, it is not necessary to feed the fish.

You wonder why fish are not fed in the winter? The reason is that their metabolism shuts down to the point that they need little food. The fish look for a warm place at the bottom of the pond, where they literally become rigid. This is where the term winter rigidity comes from. During this time they move very little and therefore do not need food.

Welcome the spring

When winter comes to an end and the temperatures increase, it is time for the fish which stayed in the house during the winter to be moved back into the pond. Those fish which stayed in the pond during the winter can now gradually be accustomed to high-quality food again.

When do fish come out of winter torpor?

As soon as the temperatures reach about ten degrees Celsius, the fish slowly start to wake up. When the thermometer reaches twelve degrees Celsius, it is also time to feed the fish again, as they now become more active and need strength. Eight weeks later, when the temperatures are at 15 degrees Celsius, it is time for the normal amount of food to be given.

TIPS FOR WINTERIZING THE POND

Here are a few more important tips so that winterizing the pond does not end in disaster:

1) The surface of the pond should not freeze completely, even in winter, for optimal oxygenation.
2) The circulation pump and filter system should be switched off to prevent freezing to the bottom. An aerator stone is optimal here.
3) In case of severe frost, an additional pond heater can be used.
4) The alternative to the aerator stone is to pour hot water at the edges of the pond to create holes. Through these, oxygen can then enter.
5) Keep the holes free of leaves and make sure the ice surface is never completely covered with snow.

Important: Do not chop holes in the ice layer, otherwise the fish will feel these vibrations and be disturbed in their rest. Likewise, you should not walk on the ice surface, as the fish would also feel these vibrations.

If your fish spend the winter in a rain barrel, it is also important to leave them alone.

- Chapter 10 -

VARIOUS GOLDFISH UNDER THE MAGNIFYING GLASS

CARASSIUS GIBELIO FORMA AURATUSA

PROFILE:

Name	Goldfish (Carassius Gibelio Forma Auratusa)
Body size	Up to 35 cm
Food	Aquatic animals, plants, spawn

Spending winter in the pond	Is possible
Keeping	Easy
Shoal size	At least 3 to 5 fish
Swimming area	Close to the surface
Water temperature	10 to 20 °C
Keep together with(Always talk to a professional in this regard).	Minnow, bitterling, koi, sunbleak

The Carassius Gibelio Forma Auratusa is a very peaceful fish. So if at some point you decide to add other species, it will not be a problem for him. We think of him as the good soul among all ornamental fish, because he gets along with almost everyone.

It is important that you always buy at least three fish, because this species in particular loves company and needs it. Leaving them in the pond is not a problem as long as the water depth is at least one meter. Carassius Gibelio Forma Auratusa prefers to stay in the lower part of the pond at that time, because it is comfortable there in winter.

Remember that it needs a water temperature of 20 °C in summer. This should not be exceeded in any case. Therefore, it is highly recommended to place the pond so that it is in partial shade.

GOLDFISH BLACK MOOR (CARASSIUS AURATUS)

PROFILE:

Name	Goldfish Black Moor (Carassius auratus)
Body size	Up to 35 cm
Food	Live food, Frozen food, Granulated food, Flake food
Spending winter in the pond	Is possible with a pond depth of at least 100 cm
Keeping	Easy
Shoal size	At least 5 fish
Swimming area	Close to the surface

Water temperature	4 to 24° Celsius
Keep together with (Always talk to a professional in this regard).	Goldfish

Black Moor is one of the most popular pond fish species, but also one of the most popular goldfish species. Like all goldfish, it is a freshwater fish. It is a breeding form of the Prussian carp, which is especially widespread in Asia. As the name suggests, we recognize it by its dark coloring. In addition, it is spherical and has very distinctive eyes.

Those who want to get this fish should know that it should be kept in the pond. It is important that the pond holds at least 500 liters of water, because the fish can grow very long and at least five fish should be kept together. In order for the fish to remain in the pond during the winter, a pond depth of at least 100 cm is necessary. The Black Moor is a very sociable and peaceful fish. Therefore, if you want to keep this one, you should definitely buy five fish. However, it does not matter what color goldfish you buy.

Anyone wishing to keep the Black Moor will find feeding it easy, as it tends to eat almost anything. However, its favorite foods are mosquito larvae and water fleas. It is important that you feed it a varied diet, suitable for a dark goldfish type.

SHUBUNKIN (CARASSIUS AURATUS)

PROFILE:

Name	Shubunkin (Carassius auratus)
Body size	15 - 30 cm
Food	Omnivore : plants, live food, frozen food, tablets, flakes
Spending winter in the pond	Is possible with a pond depth of at least 100 cm
Keeping	Easy
Shoal size	At least 5 fish

SARASA COMET (CARASSIUS AURATUS)

PROFILE:

Name	Sarasa Comet (Carassius auratus)
Body size	Up to 30 cm
Food	Omnivore : plants, live food, frozen food, tablets, flakes
Spending winter in the pond	Is possible with a pond depth of at least 100 cm
Keeping	Easy
Shoal size	At least 5 fish

Swimming area	Close to the surface (at night) At the bottom of the pond (daytime)
Water temperature	4 to 20 °C
Keep together with (Always talk to a professional in this regard).	Goldfish

The body as well as the coloring are similar to that of the goldfish. As with the Shubunkin, the tail fin is the special feature of this goldfish species. It is deeply forked and very long. The fin ends are pointed. In addition, it does not grow larger than a normal goldfish.

It is also one of the more sociable and peaceful omnivores among goldfish. Therefore, it is not a problem to keep this fish. Among its favorite foods are worms, insects and small aquatic animals. The company of other fish is very important for it. Therefore, it is recommended to keep it with at least four other fish. During the day it prefers to stay in the deep regions of the pond. However, in the evening it likes to come to the surface to enjoy the small mosquitoes.

This is a cold water fish. Therefore, the water temperature should not exceed 20 °C. In addition, it should be provided with many shaded areas. Spending the winter in the pond is not a problem, but it is important that the pond does not freeze right down to the bottom. Those who cannot ensure this, should put the fish in a tub or barrel, which can be in a quiet, dark and safe place (preferably in your own basement).

VEILTAIL (CARASSIUS AURATUS)

PROFILE:

Name	Veiltail (Carassius auratus)
Body size	Up to 25 cm (but usually only 10 - 15 cm)
Food	Omnivore : plants, live food, frozen food, tablets, flakes
Spending winter in the pond	Only indoors at 12-15 °C
Keeping	There are many things to consider here (see text)
Shoal size	At least 5 fish

Swimming area	Under the surface
Water temperature	12 to 25 °C
Keep together with (Always talk to a professional in this regard).	Goldfish, other cold water fish

Unlike the species described so far, the veiltail is very sensitive to cold. In addition, it is necessary that the water is filtered extremely thoroughly. The veiltail is also a breeding form of the goldfish. Originally it was bred in China. As its name suggests, it is the tail fin that gives it a wonderful appearance, but its dorsal and anal fin are also special. Its body is round and it usually grows up to 15 cm (but in exceptional cases up to 25 cm).

The coloration of the fish can range from red, orange, white to black. However, there are the so-called fritillaries among the veiltails. These can be two to three colored. Some veiltails also have glistening scales. Therefore they carry the name: “Pearl scale”.

Like its predecessors, it loves the company of other fish. It needs them to feel completely comfortable. So, if you should acquire a veiltail, it is highly recommended to buy at least five. It is also one of the peaceful fish species.

In summer, the veiltail can be kept very happily in a pond. The prerequisite is that it is at least 80 cm deep and holds about 500 l of water. A favorite occupation of the veiltail is burrowing, so you should also make sure that the bottom is soft. Pond soil is wonderful for this. It also loves to be able to hide. However, since it also likes to eat plants, you should select fast-growing plants here. It is important that you make sure that the water is well filtered, as it reacts very sensitively to cloudy water.

Due to its body shape, the veiltail is very fragile. Therefore, you should not let it stay in the pond during the winter under any circumstances. However, it is important for this fish to stay in a cool place during the winter, otherwise its metabolism has to work very hard. The optimal temperature of the water is 12 to 15 °C. Like the other gold fish species, the veiltail likes to eat everything. Therefore, you can offer him a varied diet without hesitation.

WHICH GOLDFISH SPECIES ARE EASIER TO CARE FOR?

A goldfish is beautiful to look at and an eye-catcher in any garden pond. But what many people should definitely do before buying a goldfish is to inform themselves sufficiently about the fish and their care. Of course, you have already learned something about that in this book , but the advice of a professional can give you much more important and also interesting information.

So that you know what you need to deal with before you get goldfish, I have created a small checklist for you here, which you are very welcome to use.

This is what goldfish keeping is all about:

- Obtaining the right accessories for care
- Choosing the right type of food
- Deciding whether to spend the winter in the pond or aquarium
- Choosing suitable pond plants/aquarium plants
- •Planning for a sufficient number of fish
- Choosing the right combination with other fish
- Gaining a knowledge of reproduction
- Ensuring sufficient space for all fish
- Knowing how to bury a goldfish

Now let's get back to the question that the chapter mainly deals with. Which goldfish species are actually easier to care for? The goldfish, as we know it in this country, is very robust and easy to care for. This small, sociable guy is already satisfied if you give him a sufficiently large home and buy one or the other companion in addition. All goldfish species available for purchase here love company.

If you want to keep a special species, it is important that you study it in detail in advance and also get proper advice when buying. If you want to buy a breeding fish, you will certainly be able to do so at a breeder. Make sure that he can certify the papers of the fish and its affiliation to a club. This is the only way to be on the safe side and not run the risk of falling for a scammer.

WHICH GOLDFISH SPECIES ARE MORE SENSITIVE?

As you have already learned, there are many different breeding forms of the popular goldfish. But can all of them actually be kept in a garden pond or are there also species among them that are more vulnerable? The answer is: Yes, there are!

A veiltail, for example, is very sensitive to cold temperatures. With this fish, care must be taken to stick to the care instructions given, such as that the temperature of the water, which should not fall below 12 degrees Celsius, and that it should not spend the winter in the pond under any circumstances. The reason for this is its round body shape.

Some fish are very heat-dependent and thus not as hardy as the normal goldfish. Others have very bizarre and unnatural fins. Their different body shapes are also a reason why some species are not intended for keeping in the pond.

These species are not suitable for keeping in the pond:

- Veiltails (summer only)
- Telescope eye
- Double Veil Fish
- Stargazer
- Buffalo fish
- Eggfish

- Lion heads

For example, if you are considering buying one of the above fish, you should find out exactly what the fish likes best and how it can be kept in a suitable environment. Note: There are also many breeds that are bred under torturous conditions. It is therefore forbidden in Germany to sell these in specialized shops.

ARE THERE DANGEROUS GOLDFISH?

A goldfish is supposed to be dangerous? No way! Or is it? As bizarre as this may sound, it is actually true! However, we humans are to blame. The fish does not threaten us, but those amphibian species that are very rare and live in wild waters. How can the goldfish become dangerous to them? There are actually people who acquire goldfish, but eventually realize that they no longer feel like taking care of them. Instead of passing them on, some people simply release the fish into the wild.

If you have followed the history of the goldfish, you already know that the fish has always been a pet. If fish are left to fend for themselves, they can cause certain problems, such as disturbing the ecological balance of the water. You are wondering how such a small fish can make such a difference to the ecological balance? The reason is that it is greedy and will anything that gets in its way, including spawn or tadpoles.

If a goldfish is released into a pond, it can eat half of the small living creatures there. This means that an entire population of frogs and other amphibians could be endangered. Of course, this is fatal for the protection of rare species.

The goldfish is an intruder, also called a neozoon, in a wild body of water. Even though it is very adaptable, that does not mean it is not dangerous. Many abandoned fish do die, but some are very resilient. These have a strong will to survive. If they reproduce - and we have already learned that the female can lay up to 3000 eggs - soon the whole habitat will be populated by goldfish, while other species will become fewer and fewer.

The goldfish is a freshwater fish, but it has been found that it can even live in salty, brackish water. Meanwhile, it is found in wild waters all over the world. We can therefore state that the goldfish is harmless to humans, but not to its environment if released. Therefore, think carefully about keeping goldfish and never release your fish into the wild!

- Chapter 11 -

ADDING THE GOLDFISH

Now that you have purchased your goldfish, they are finally going home with you. But before they can be put into the aquarium or pond, there are still a few things to consider. If many mistakes are made, it means stress for the fish and in the worst case can even lead to their death. For example, the transport of the fish can cause a lot of stress.

First, let's look at how best to capture goldfish. Transporting a fish can also become necessary when you move to a new apartment or house. So that you are well prepared , I have compiled the most important things for you to bear in mind.

Before a fish can be transported, it must of course be caught. We do this with a dip net. It is important that this is at least four times the size of the fish. By the way, it is best to choose two dip nets, one for guiding the fish and the other for catching it. Once the fish is in the net, lift the dip net out of the water and grab it with one hand underneath. This avoids unnecessary wriggling of the fish, which would only cause it stress. For this reason, it is also recommended to work in pairs. This way, one of you can take full care of the captured fish and put it into the transport container. It is very important that you do not leave the fish out of the water for too long, so prepare everything very carefully in advance.

Once all the fish are in the transport container, you should proceed directly with the transport. When choosing the container you must be sure to observe the following things:

- The container must be 100% waterproof,
- it must be adapted to the size of the fish and
- be filled with ⅓ water and ⅔ air.
- For longer journeys, transport inside a polystyrene box
- Transparent containers should be darkened
- In case of much stress, a sedative may also be administered

PUTTING IN THE AQUARIUM

When you arrive at your destination, the fish have already survived the worst of the stress. If you think that the fish can now be put directly into the aquarium, you are mistaken. The water is completely different from the one in which they lived before. Therefore, it is necessary that you first accustom your fish to the aquarium water.

If the fish are in an air bag, it can simply be placed on the water surface of the aquarium. It is important that it remains closed. Why it must be done this way? In this way you can adjust the water temperature in the air bag to that of the aquarium. In addition, this way the fish can look at the new environment quite well. It is recommended to let the bag float on the water for about 15 minutes.

Now it gets exciting and your full concentration is required! Now open the bag with the edges facing downward. Pour a little of the water into the aquarium every ten minutes. This allows the fish to get used to their new environment.

After about 30 to 60 minutes, you may now let the fish into the tank. A second person catches them with the dip net and then carefully lets them slide into the aquarium.

Tip: Prepare the aquarium about three weeks before you buy your first fish. This gives the water enough time to stabilize, which is important because only then will the tank be ready for the fish. Of course, there are special agents available in specialized stores that accelerate the stabilization process, but I do not recommend this, especially if you are getting fish for the first time.

Quarantine the fish:

If you already have fish in the aquarium and buy new ones, it is highly recommended to put the fish in a separate tank first. The reason for this is that the fish can carry bacteria or even diseases with them. They would then transmit these to the fish already present. I am sure you want to avoid this. Once the new ones are in quarantine, you will need to watch them very closely.

Here's what to keep in mind:

- Fish behavior
- Fish appearance
- Swimming behavior
- Feeding behavior
- Possible signs of diseases

ADDING FISH INTO THE GARDEN POND

If you buy new fish for the pond, you naturally want to put them in quickly. But here, too, you should proceed as you have just read. Only in this way can you accustom the fish step by step to the new environment and the new water, as well as detect any diseases, without endangering the fish already in the pond.

When is it suitable to put fish in the pond?

It is ideal if you add the fish in the spring. It is slowly getting warmer and thus the temperature of the water also rises. Of course, it is also possible in the summer, but you should remember that the autumn could be very cold and the fish will have to be removed from their new home quite quickly. For this reason, I advise you to wait until next spring to do this.

If the pond is newly made, it is also necessary that the pond can first establish a biological balance. This, of course, takes time. You can take your cue from the plants that have been put in; these need about twelve weeks before they have become accustomed to the new environment and contribute to the establishment of the new biotope. After these twelve weeks, the fish may now also enter. The water temperature should always be around 12 degrees Celsius.

- Chapter 12 -

BREEDING FORMS

When talking about breeding forms, we are not talking about a species or breed. With goldfish it is like with dogs: There are no species, because the fish have the same ancestors and are able to reproduce among themselves. The goal of a breed is to breed purebred trait carriers. But this is not possible with goldfish.

In 1908, the German-language standard on goldfish was published. According to this standard, four forms are important for the valuation. In the English standard, which appeared in 1977, there is talk of seventeen forms. If we look to China, we find already over 450 variants.

Japanese breeds are either further-developed forms of the Chinese breeds or ones which have been developed by themselves. In fact, most of the breeding forms in Germany have Japanese names. As you can see, there is a very large variety. How are we supposed to sort them out?

In ancient China, people did not intend to breed fish that could be listed in catalogs. Their goal was to have a fish that was unique. It is Europeans who express the desire to classify fish. If we look in the pet shops, there does not seem to be a specific standard for them. Most of the fish are not classified and show varying characteristics. Added to that, they are given fantastic names to help the sale along. Therefore, we Europeans should perhaps think back to what the ancient Chinese wanted: To own a unique fish.

To give you a rough overview, I would like to introduce three breeding forms in more detail. In addition, I would also like to refer to some controversial breeding forms, as some of these are prohibited.

ELONGATED BREEDING FORMS

Common features	Special features	Keeping
• All fins are simple • Body similar to the wild form • Coiled body	• Coloration deviating from the wild form • Possibly larger or differently shaped fins	• Robust • hardy • Pond and aquarium possible • Tail fins and body size important for water quantity

Among them is the **normal goldfish**. It is the original goldfish, so to speak. You have already learned a lot about them.

The comet tail is also an elongated breeding form. Its body shape resembles that of the normal goldfish. It got its name because of its enlarged and elongated caudal fin. However, it is also usually the case that all of its fins are enlarged. The pelvic and pectoral fins tend to be pointed and have a straight edge. Americans often consider themselves the " developers" of these fish, even though they originated in Japan. If you provide enough space for a comet tail in your aquarium, and offer it very good nutrition , you will ensure that very majestic fish will develop. Pay attention to the tail fin when buying. If it has already grown very large in young fish, this is a sign that it will remain very disproportionate as it gets older.

If a comet tail has a very striking gray-brown-olive coloration, it is called an iron comet. These are said to be the result of a cross between a fringe tail and a Prussian Carp. These fish are very rare. Sometimes there are black fish swimming around in a shop tank. We always get a surprise with those, no matter if they are cheap or expensive to buy. They do not keep their black color. They change color much later. In some cases they even lose their red pigment. The permanent wild coloration is rare. In fact, they are more likely to be found in the offspring of comet tails that live in the pond.

Another elongated form is the **shubunkin.** This one has a speckled calico coloration and is also mistakenly called scaleless. If we look at the British standard, the London Shubunkin is described as being very similar to the Common Goldfish. Only the surface is different, namely not metallic, but with a mother-of-pearl shimmer in all colors and combinations. The Bristol Shubunkin has a lobed, enlarged caudal fin with rounded ends and a high dorsal fin. The word shubunkin does not refer to the color of the fish. However, the Japanese use it to mean the long-finned fish. The American Shubunkin also follows this tradition. If we look in our pet stores, we will find numerous shubunkins that do not meet any of these standards. The only characteristic is their calico coloration.

Then there is the **grass goldfish**. What is special about it is that it is elongated, but has a split caudal fin, which is carp-like in size, but can be very variable in shape. In some fish the caudal fin is completely divided, while in others it is divided only half-way. The latter resembles the tail of an airplane. The trunk as well as the other fins resemble the normal goldfish. In China, these fish are considered the ancestors of the veiltails, which I will introduce later.

BREEDING FORMS WITHOUT DORSAL FINS

Common features	Special features	Keeping
• Missing dorsal fin • Compressed body (egg fish) • Elongated torso (protruding eye)	• Coloring diverse	• Only suitable for aquarium

The **eggfish** is one of these forms. Its appearance is stocky and compressed. In addition, it usually has short fins, a divided caudal and no dorsal. This fish is an old form and has become very rare.

The Lionhead is also a fish without a dorsal fin. Its caudal fin is very short and almost vertical. Its back is also horizontal and goes into the tail part. It is seen as an image of the guardian lion of imperial China.

The **Japanese buffalo head**, also called **ranchu**, has a round body shape and a tail portion which is tilted at 90 degrees. The caudal is divided and carried low. In addition, this is short and splayed. It also has a head cap, but it is less pronounced compared to the lion's head. It goes down far on the sides of the head.

It is one of the top specimens among Japanese breeders, which makes it highly prized. Its breeding is extremely difficult and even specialists achieve only 40% success. Due to its cuddly appearance it is very popular with beginners. However, especially at the beginning it is not advisable to buy this fish. In our pet shops they are available, but they easily fall ill, which quickly causes the owner to be overwhelmed.

The **stargazers** owe their name to their upward pointing eyes. Just imagine how strange these fish must look when we look at them sideways in the aquarium. As a result, they were bred more for top-viewing in tubs. Americans consider this fish to be very hardy, which is why it usually finds a home in their ponds.

The **bubble eye** owes its name to the special feature it carries under its eyes: Bubbles. It is a relatively slender fish and usually has a long, divided caudal. In Asia, you can also find stargazers that have double bubbles under their eyes or variations with smaller bubbles. These are called toadheads. However, it is not advisable to buy these fish, as they very much meet the criteria of animal cruelty.

VEILTAILS

Common features	Special features	Keeping
• Compressed, stocky torso • Varying height back • Split tail fin	• Coloring diverse • Strongly enlarged or differently shaped fins (true veiltail: long and drooping) • For others short and erect • Possibly thickened eyes • Head growth with some breeders • Pearl scales • upturned gill cover • Partially pompon like,	• needs heat • challenging • Vulnerable

	fleshy growths on the nasal septum	

This form includes, for example, the **Ryukin**. Its body has an egg-shaped or hump-backed form. The caudal fin is long and divided and complements the anal fin, which is also divided. For this reason, this fish is also described as "four-tailed". This form originated during the Ming period. On Okinawa, this form has been bred since the 17th century. The Japanese name Ryukyo kingyo, or Ryukin for short, was given to it in the 18th century.

The **fantail**, has a compressed body shape and a completely divided / forked caudal fin. Like a fan, it stands upright and is splayed. The Asian fish of this species were bred for top-viewing, so they were kept in tubs. The British and American ones, on the other hand, were bred for front view in the aquarium. Their ancestors are the grass goldfish. They used to be slender and only acquired their compressed shape in the 20th century.

The **Veiltail** has a short and rounded body shape. The caudal fin has the shape of a veil, is long and drooping. The anal fin is also divided. Dorsal, ventral and pectoral fins are also long. In addition, the veiltail holds its long dorsal fin upright.

This form is sensitive to cool water and therefore not recommended for keeping in a pond.

The dragon eye is also interesting. The thickened, spherical eyes are the decisive feature for its name. Incidentally, it is also called **telescopic eye**, because the eyes protrude in a cylindrical or conical shape. Depending on the breeding form, these can be found in different colors. By the way, dragon eyes used to be an elongated fish, but when they were crossed with Wen-shaped fish, this changed. Moreover, the slender forms are no longer bred nowadays. According to the German standard, the Telescopic Fish is pot-bellied and black with spherical eyes. In the British standard it is called **Black Moor**. But this naming is nonsense, because it is available in all possible colors.

Dutch lion head is not as Dutch as the name suggests. In the 19th century it became known in Nagasaki and was later imported from China. It was in Japan that it ultimately gained its popularity. By the way, its Japanese name is Oranda, which means Holland. Its body shape as well as fins are similar to that of the Veiltails, but they are more diverse. Therefore, they are also less strictly standardized. They have a very characteristic head growth, which can be seen in different shapes and sizes. This fish can sometimes grow up to 60 cm. Anyone thinking of buying this fish, should therefore absolutely make sure that the aquarium has a sufficient size. The Dutch lionhead is not suitable for beginners.

CONTROVERSIAL BREEDING FORMS

With this guide I would like to provide information about goldfish. However, since there are points of criticism especially on the subject of goldfish breeding, I would like to draw your attention to these as well. Some breeding systems are cause for concern from an animal welfare point of view. In particular, advanced aquarists try to bring a little nature into their own homes, but this often clashes with the natural development of the goldfish themselves. For thousands of years these fish have been revered for their diversity. What you yourself prefer should always be accompanied by the thought of whether this is in the best interests of the fish. In addition, expertise is needed, but also basic knowledge of natural sciences.

There are books from the 18th century which describe the so-called sleeper, which kept sinking to the ground or floated on its back in the water. It probably had a severe swim bladder defect. This illustrates that severe defects can occur. If we look at § 11b of the animal protection law, we quickly find out that the breeding of vertebrates is forbidden, if they have to suffer because of their characteristics or cannot behave in a way which is typical for their species.

In Sweden, for example, the sale of orandas is prohibited. In Austria, Switzerland and also here in Germany, the demand for bans is very high. The problem with that is that a lot of the criticism is aimed at individual cases, rather than there being a general problem and some criticisms make no sense at all. One example of this is that the Pearlscale goldfish is said to have genetically ruffled scales.

Often, discussions about it show great ignorance. If we look at it from the critics point of view, it even seems patronizing. Can we really assume that there are goldfish that suffer because of their breeding characteristics? Harmless phenomena are discriminated against and real problems are overlooked as a result. In fact, not all problems which a goldfish has are due to breeding traits. Therefore, their appearances are often not the problem, but the selection of which fish actually find their way into the shops. In addition, many keepers have inadequate conditions for keeping the fish.

Very often in this country goldfish breeding forms are offered at very low prices in the double-digit range, making careful breeding, which leads to healthy fish, hardly possible.

If we look at the selling prices of koi, three-digit sums are common. We pay up to four digit sums for the image. People who are not willing to pay reasonable prices do not need to wonder about "problem fish" afterwards. Critics should not be complaining about problems with the breeding forms. It is often not the breeding trait itself that is the cause of fish suffering, but the cheap breeding and the importation of those fish. A goldfish breeding program must proceed carefully, because only in this way can problematic cases be excluded.

If we look at the compressed fish, we often notice that they have problems with digestion. The reason for this is that their intestines are long, but constricted. In addition, there may also be incorrect feeding. Gas formation, swallowing air when the fish eats at the water surface, or even food that swells up will cause the fish to float. Downplayed, this is called corking. In addition, it is also possible that the posterior swim bladder chamber is atrophied, missing, or both chambers are incorrectly positioned. Sometimes fish also have inverted gill covers, which do not close properly. This interferes with the fish's ability to breathe. In some ryunkins, the gill covers abut the thickened trunk, preventing them from closing. Therefore, during periods of rest, they may suffer a lack of oxygen. Water that is oxygenated and moving is important for these fish, and yet many still suffer.

The so-called nose pompons have protrusions in front of the mouth which can interfere with the inflow of water. The fish may then suffer from respiratory distress. There is a risk of injury.

A goldfish likes to dig very much. If we look at the bubble eye, the criticism gets louder, because the bubbles are sometimes so large that the fish can't burrow. In addition, the bubbles are filled with lymph. The amount of lymph in the bubbles could amount to as much as it has throughout the rest of its body. This alone is an extreme physiological challenge for the fish. There are owners who report that their fish's organs have been compromised due to the high amount of lymph in their bladders. Fish which have very large bladders also get sick more often.

However, there are also studies which show that regeneration of the bubble eye goldfish is enormous. Researchers hope to be able to use this medically. There is a need for a study, only then can a clear statement be made about it.

Some fish have a head growth which can be extreme, covering the eyes of the fish and hindering them in their movement. Often the affected fish suffer greatly and lie around on the bottom, causing sores to their bodies.

So, although these fish are special and often look beautiful, it is probably wise to steer clear of them and instead stick to breeding forms where the welfare of the fish comes first.

- Chapter 13 -

ARE GOLDFISH EDIBLE?

"But while our affection for fish is otherwise predominantly determined by the pleasant taste, few people have thought of eating goldfish as well." (Grzimek, 1970)

If you expect recipes in this chapter, with which you can conjure up something wonderful for your goldfish, then unfortunately you will look in vain here. It is actually the case that you can eat goldfish. Unbelievable, isn't it? So that you believe me, I present you two recipes with which you can actually prepare the goldfish. For most pet owners, it would probably never occur to them to eat their own pets. I can't bring myself to do that with my fish either. Nevertheless, I do not want to withhold this information from you, even if it might seem strange to you.

In Asia, for example, the goldfish or their wild form have been eaten for many millennia. At the present time, in fact, the goldfish are still eaten.

Those who want to eat the fish from their own garden pond, but want to avoid the swampy taste, should keep the goldfish in clear water for a few days. However, fish that are sick, have received medication or died naturally should not be eaten.

GOLDFISH WITH BEAN SAUCE (SPICY)

Ingredients:

- 1 goldfish (approx. 150 g, live)
- 2 tbsp soy sauce
- 1 tablespoon rice wine
- 2 tsp salt
- 1 cup water
- ½ cup oil
- 2 tablespoons chopped green onions
- 1 tablespoon chopped ginger
- ½ tbsp vinegar
- ½ tablespoon cornstarch
- 1 tablespoon chopped garlic
- 2 tablespoons hot bean sauce (doubanjiang)
- 1 tsp sugar
- 1 tsp sesame oil

Preparation:

1) The fish must first be killed and gutted. Then cut it on both sides into 3 strips.
2) Now fry them in oil.
3) Put the strips aside and use the remaining oil to sauté the ginger and garlic.
4) Now add the hot bean sauce.

5) When everything smells nice, the rice wine is added along with the soy sauce.
6) Mix everything with sugar and salt.
7) Now add a little more water. Place the fish in the middle.
8) Let it cook for three minutes and keep basting the fish with the sauce.
9) Stir some cornstarch with water. With this mixture you can bind the sauce. Also add a little vinegar and sesame oil.
10) Serve the fish on a deep plate with rice and spring onions.

GOLDFISH SWEET AND SOUR

Ingredients:

- 1 large goldfish (very fresh, approx. 600 g)
- 1½ tsp salt
- 4 tablespoons black vinegar
- 3 tablespoons rice starch
- 4 tablespoons water
- 2 spring onions
- A little pepper
- 2 tablespoons soy sauce
- ½ tablespoon sesame oil
- 4 tablespoons lard or oil
- 2½ cups clear broth
- ½ cup fresh ginger in strips
- 2 slices ginger
- 3 tablespoons sugar
- 1 tablespoon rice wine or Shaoxing wine

Preparation:

1) Kill the fish, gut it and clean it.
2) Cut it open at the belly. It should still be attached to the backbone.
3) Smash the backbone in several places. Use a blunt knife for this.
4) Place water, ginger slices, and scallions in a large pot and bring mixture to a boil.

5) Place the fish in the water with the outside facing up. Let it cook like this for two minutes.
6) When the eyes turn white, plate the fish with the outer side upward.
7) Soak the ginger slices in cold water, take them out and squeeze them dry. Together with pepper, sprinkle them over the fish.
8) Heat 3 tablespoons of lard and pour in the rice wine along with the clear broth. Also add all other ingredients except rice starch and sesame oil. Bring the mixture to a boil.
9) Heat the lard with the sesame oil and pour this over the fish with the sauce.

- Chapter 14 -

TRUE OR FALSE? THE MOST COMMON MYTHS ABOUT THE GOLDFISH

Can goldfish burst?

The goldfish in the pond or aquarium is getting fatter and you are afraid that it might burst. This conjecture is circulating on the internet , but you are not sure if there is any truth to it.

Well, I can tell you this. A fish that is getting fatter and fatter is either getting too much food or it is sick! Therefore, control your feeding and consult a fish veterinarian to examine your fish. A fish cannot burst in that sense, but if it is fed too much, for example, there is a risk that it will suffer considerable injury and its lifespan will be reduced.

The goldfish has a bad memory

Do goldfish even have a memory? I can hear this question buzzing through your head at right now. Goldfish, or fish in general, are said to have only a three-second memory. But this is completely wrong! Goldfish can even remember something for up to five months!

A goldfish can be made drunk

A few drops of beer or wine won't hurt a fish, will it? Goldfish can actually get drunk! Alcohol has a different effect on the little creature than it does on us. For them, just a few percent is enough. Researchers experimented with zebrafish, which they let swim in alcohol. The zebrafish suddenly swam faster and behaved more confidently.

A goldfish survives even in milk

I cannot give a definite answer here, but here is a brief explanation: As long as there is oxygen in the milk, this is probably possible. However, if this dwindles or is used up, it becomes unpleasant for the fish. So in this sense the chances of survival very low for the poor fish, so I would not recommend this under any circumstances!

Goldfish sleep with their eyes open

When a fish sleeps, it slows down and sinks to the bottom. But since a fish does not have eyelids like we do, it cannot close its eyes. So it is indeed true that goldfish sleep with their eyes open.

Goldfish have no feelings

That is wrong. A goldfish also mourns the death of its companions, but in its own way. Fish even feel stress, fear and pain. So that is pure ignorance that has lasted for many years.

CLOSING WORDS

Yes, goldfish can be wonderful ornaments. If you want to give them a pleasant life, as you may have already noticed while reading, it is necessary to get enough information. It does not matter that goldfish are quite robust and easy to please. Without expertise and important information, you cannot keep fish in an appropriate way.

As a little bonus, I'd like to finish by giving you a checklist that you can use as a guide when getting that aquarium or making your backyard pond.

I hope that I have been able to provide you with extensive and interesting information about our popular ornamental fish with my guidebook and wish you good success as well as lots of joy with your aquarium or with your pond as well as its future inhabitants.

CHECKLISTS

FOR THE PURCHASE OF AN AQUARIUM

- ❑ Aquarium tank
- ❑ Window cleaner
- ❑ Back wall
- ❑ Base for the tank
- ❑ Control heater with thermostat (25-30 watts)
- ❑ Thermometer
- ❑ Filter, filter starter, filter material
- ❑ Depot fertilizer for the substrate
- ❑ Decoration: stones / roots
- ❑ dip net
- ❑ Timer for the lighting
- ❑ Hose 1.5 m for water change
- ❑ Water test set
- ❑ Approx. 25 kg aquarium gravel (grain size 3-5 mm)

- ❑ Stable base cabinet
- ❑ Aquarium cover with integrated lighting
- ❑ 10 l bucket (use only for aquarium!)
- ❑ Water conditioner
- ❑ Plants

FOR POND CONSTRUCTION

- ❑ Basin
- ❑ Excavation
- ❑ Sand
- ❑ Water
- ❑ Plants
- ❑ Fleece
- ❑ foil
- ❑ Gravel
- ❑ Decoration
- ❑ Pump
- ❑ Filter
- ❑ gravel cleaner
- ❑ dip net
- ❑ Thermometer

FOR THE CARE OF THE POND IN THE SUMMER

- ❑ Fish out thread algae regularly
- ❑ Remove leaves
- ❑ Clean pond pump and filter
- ❑ Check pond pump and filter
- ❑ Check water temperatures
- ❑ Check water values
- ❑ Check water quantity
- ❑ Prune the Plants and vegetation at the edges
- ❑ Pay attention to fish stocks
- ❑ Check pond bottom
- ❑ Feed with caution

POND CARE AND PREPARATION FOR THE WINTER

- ❑ Remove dead plant remains in fall
- ❑ Remove leaves and algae remains with gravel cleaner
- ❑ Leave winter hardy plants in the pond
- ❑ Cut back water lilies vigorously
- ❑ Bring tender plants indoors during the winter
- ❑ Take out tender pond plants before the first frost
- ❑ Install protection net over pond
- ❑ Fill completely with water by the beginning of November at the latest

Without fish stocking:

- ❑ Take technical devices out
- ❑ Turn off the pond pump and take out at temperatures below ten degrees
- ❑ Drain pump and store in a frost-free place
- ❑ Never leave the pump dry during the winter

- ❑ Switch off pumps outside the pond and protect them from frost
- ❑ Dismantle UVC pre-cleaner in good time, clean it and also store it frost-free.
- ❑ Outdoor filter box: rainwater must not be allowed to penetrate
- ❑ Completely empty and clean the filter box, hoses and pipelines.
- ❑ Remove and clean foam inserts and substrate tubes

Prepare ice free holder:

- ❑ Weight air chambers with sand (Use with pump (alternative))
- ❑ Use pond balls alternatively

With fish stock:

- ❑ Feed fish for the last time in autumn
- ❑ Pond depth below 80 cm: remove fish from the pond in autumn
- ❑ Never break the ice
- ❑ Pond heater for ornamental fish
- ❑ No longer feed in winter
- ❑ Feed in a heated pond in winter

BOOK RECOMMENDATIONS

Get the first part now and learn how to train your Labrador puppy!

Labrador Training – Dog Training for your Labrador puppy

Dog training is often:

"... confused with classical dog training

"... considered necessary only for demanding dogs.

"... considered too difficult to manage without experience.

But what does dog training really mean and what is it good for? And how can your Labrador and you benefit from it even without any experience?

The most important thing is to first understand how a dog perceives its environment, what is "normal" for it and how you can use that to your advantage. In addition, the characteristics of each breed are crucial when it comes to later training. Your Labrador, for example, has different character traits than a Dachshund, and it is precisely these differences that must be taken into account in dog training.

Read about background information for testimonials, step-by-step instructions and insider tips tailored to your Labrador.

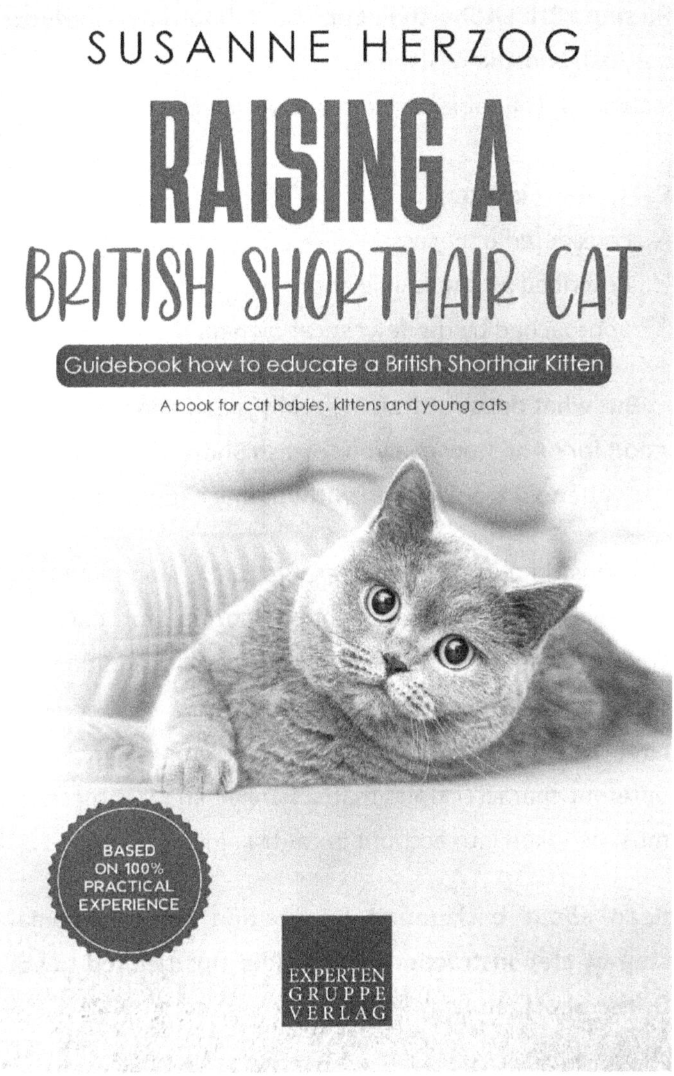

Get the first part now and learn how to train your British Shorthair cat!

Raising a British Shorthair Cat – Guidebook how to educate a British Shorthair Kitten

A book for cat babies, kittens and young cats

Cat training is often...

"... considered unnecessary,

"... described as impossible and

"... approached by the fewest cat owners

But what does cat training really mean and what is it good for? And how can your British Shorthair cat and you benefit from it even without any experience?

The most important thing is to first understand how a cat perceives its environment, what is "normal" for it and how you can use that to your advantage. In addition, the characteristics of each breed are crucial when it comes to later training. Your British Shorthair, for example, has different character traits than a Bengal, and it is these that must be taken into account in cat training.

Read about background information for testimonials, step-by-step instructions and insider tips tailored to your British Shorthair.

PERMACULTURE
FOR BEGINNERS
KNOWLEDGE AND BASICS
OF PERMACULTURE
100% EXPERT
KNOWLEDGE
MADE IN
GERMANY
FRIEDHELM WEISS

Permaculture for Beginners – Knowledge and Basics of Permaculture

Wouldn't it be great if our buildings and vehicles could filter water while also improving air quality? With this intention to replenish the environment, our world could be wonderfully clean.

But how exactly is permaculture changing the world?

For one thing, permaculture can transform deserts into productive oases for food production. For another, it can bring us the much-needed solutions that can reduce, if not reverse, the human impact on our planet that is so obviously harmful today.

Everything is subordinated to the one goal of creating permanently and sustainably functioning cycles that run in harmony with nature.

Here are the main permaculture philosophies explained to give you a broad insight into the field of permaculture

Start today and go ahead to save our planet with a little more permaculture!

DID YOU LIKE MY BOOK?

You have read my book and now know everything there is to consider when keeping goldfish in your own aquarium or pond. And that is exactly why I now ask you for a small favor. Reviews are an important part of every product offered on Amazon. It is one of the first things that customers look at and not infrequently the reviews later give the decisive decision to buy a product or not. Considering the endlessly large selection of Amazon, this factor is becoming increasingly important.

If you liked my book, I would be more than grateful if you would leave me a review. Just write briefly what you particularly liked or how I can perhaps make the book even better. It won't take more than 2 minutes, I promise! You can be sure that I personally read every review, because it helps me a lot to make my books even better and to adapt them exactly to your wishes.

Therefore, I say to you:

THANK YOU VERY MUCH!

Yours Svenja Nett

REFERENCES

Die Sympathie der Goldfische: neue Erzählungen aus Südkorea, 1. Aufl., hrsg. von Bertulies, Friedhelm ¬[Hrsg.], unter Beteiligung von I, Mun yeol, 2005

Gesundheit fur Zierfische: Parasiten erkennen und bekampfen, von Schmahl, Gunter Mehlhorn, Birgit Mehlhorn, Heinz, 2013

Zierfische, Aquarienpflanzen, Terrarientiere: Kölner Aquarianertage '96, hrsg. von Meidling, Werner ¬[Red.], unter Beteiligung von Verein Kölner Aquarien- und Terrarienfreunde Gegründet 1946 Kölner Aquarianertage 1996, 1996

Züchterkniffe. 6: praktische Winke zur Pflege und Zucht beliebter Zierfische, 1. Aufl., unter Beteiligung von Nachstedt, Julius Gremblewski, Otto Tusche, Hans, 1984

Aquarien-Atlas., 4. Aufl., unter Beteiligung von Baensch, Hans A Riehl, Rüdiger, 1984

Fremdländische Zierfische : Winke zur Beobachtung, Pflege und Zucht der Makropoden, Guramis, Gold-, Teleskop-, Hundsfische u.a. ; nebst Anhang: Bemerkungen über die Axolotl / nach..., von Dürigen, Bruno, 2009

Die tierhygienischen Grundlagen der Süsswasseraquaristik sowie Untersuchungen über ihre Beachtung in der Zierfischhaltung, von Etscheidt, Jutta, 1990

Zierfische : Japan, das Land der Koi – Zucht, von Riedl, G, 1996

Nano-Aquaristik, von Kai Alexander, Quante, 2010

Süßwasseraquarien: Die 20 besten Unterwasserwelten, von Claus, Schaefer, 2015

Das grosse Lexikon der Aquaristik: Süsswasser u. Meerwasser, 3. Aufl., von Frey, Hans, 1983

Handbuch der Meeres-Aquaristik. 3, Knochenfische. T. 3, Knorpelfische u. Wirbellose / Wolfgang Klausewitz; Peter Wilkens: Seewasserfische und Wirbellose, von Klausewitz, Wolfgang Wilkens, Peter, 1979

Das Süßwasser-Aquarium: der bewährte Aquaristik-Ratgeber für jedermann, 27. Aufl., von Frey, Hans, 1986

Das große Buch der Meeresaquaristik, von Probst, Karl Lange, Jürgen, 1975

Enzyklopädie der Aquaristik und speziellen Ichthyologie, hrsg. von Sterba, Günther ¬[Hrsg.], 1978

Der Goldfisch - ein Haustier aus China, von Philipsen, Mechthild erschienen in Unterricht Biologie (1986), 1986 https://www.gartenteich-ratgeber.com/tiere/teichfische/arten/goldfisch/

https://www.zooplus.de/magazin/aquaristik/fischarten-portraits/goldfisch

https://www.gartendialog.de/goldfische-haltung/

https://www.kois.de/ratgeber/der-goldfisch-im-gartenteich-lebensweise-und-haltungsbedingungen/

https://aquaristik-profi.com/fische/arten/sonstige/goldfisch/

https://www.gartenlexikon.de/goldfische-haltung/

https://www.kois.de/ratgeber/der-goldfisch-im-gartenteich-lebensweise-und-haltungsbedingungen/

https://aquaristik-profi.com/fische/arten/sonstige/goldfisch/

https://www.gartenlexikon.de/goldfische-haltung/

https://aquaristik-profi.com/fische/einsetzen/

https://www.obi.de/magazin/garten/teich/fische-in-teich-setzen

Imprint

1st edition

 Publisher: GbR, Martin Seidel and Corinna Krupp, Bachstraße 37, 53498 Bad Breisig, e-mail: info@expertengruppeverlag.de, cover photo: www.depositphotos.com. All contents represented here serve exclusively the neutral information. They do not represent any recommendation or advertisement of the described or mentioned methods. This book makes no claim to completeness, nor can the timeliness and accuracy of the information presented here be guaranteed. This book in no way replaces the professional advice and care of a veterinarian or animal psychologist. The author and publishers assume no liability for any inconvenience or damage resulting from the application of the information presented herein.

Made in United States
North Haven, CT
10 November 2026

22091259R00114